The Human Side of Construction

The Human Side of Construction

How to Ensure a Successful, Sustainable, and Profitable Career as an AEC Professional

Second Edition

Angelo Suntres

Edition History
Angelo Suntres (1e, 2023)

Published by John Wiley & Sons, Inc., Hoboken, New Jersey.
Published simultaneously in Canada.

Library of Congress Cataloging-in-Publication Data Applied for:
Paperback: 9781394266203

Cover Design: Wiley
Cover Image: © PT STOCK/Getty Images

Set in 9.5/12.5pt STIXTwoText by Straive, Pondicherry, India

SKY10086352_093024

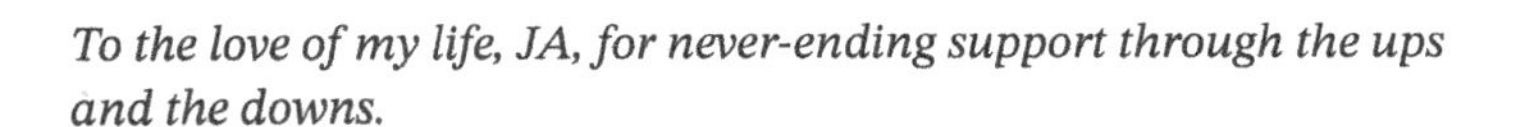

To the love of my life, JA, for never-ending support through the ups and the downs.

To my parents for making me believe that I can achieve anything.

And to anyone who has helped to shape my experience, positive or negative, thank you.

Contents

About the Author

Angelo Suntres is a passionate leader in the construction industry with 20 years of experience in designing and building in all sectors. He has represented both contractors and owners and understands the struggles from both sides of the contract. He believes that focusing on the human principles of connection including effective communication and fostering healthy relationships is a critical part to ensuring a successful and inclusive future for the construction industry.

Introduction

The construction industry is undergoing major changes driven by technology, innovation, and a new way of working including different aspirations and motivations for younger workers. These factors present challenges, which are not adequately addressed by traditional tools, methods, and processes. It will take a fundamental shift in the way we think and operate to ensure a successful, sustainable, and profitable future. This future must be focused on leadership guided by the human principles of connection including effective communication and fostering healthy interpersonal relationships.

These principles have seemingly slipped by the wayside as the industry has become increasingly combative, transactional, and litigious. Projects are becoming more complex, schedules are being compressed, reliance on technology is increasing, and contracts are becoming more onerous. And yet, it seems that as humans, we keep getting further and further apart. Adding on the complexities posed by the current talent shortage, imminent retirement of the experienced "baby boomer" generation, and the mountain of work we are all facing, the key to success in the future will be doing more with less.

With the deterioration of human connection and communication, we can observe the following issues with the industry as it is today:

- Severe lack of trust and collaboration between:
  - Owners and contractors
  - Contractors and subcontractors
  - Subcontractors and vendors
  - Employers and employees

- Prioritization of immediate financial goals over relationships and organizational resiliency
- Immense rework needed due to miscommunication or improper data
- Time wasted on nonproductive tasks like collecting project information, conflict resolution, and mistakes
- Decreased productivity due to poor site logistics and coordination
- Many more

Nevertheless, construction remains booming and vibrant propelled by population growth and the advancement of society with a cost projection of an estimated $14.4 trillion globally by 2030 compared to $7.28 trillion in 2021. As the demand and cost of construction rise, so too do the effects of widespread issues stemming from poor leadership and communication increasing strain on the industry and the people in it. This will only perpetuate talent acquisition issues, lost profit, and employee burnout if not properly addressed.

The need for human connection and effective leadership at all levels of organizations has never been greater; the time is now.

This book was created to solve these problems, overcoming barriers through personal development skills that provide a positive professional impact for individuals, organizations, and the entire industry. It is for people who have a passion for leadership and want to further their careers while adding value to those around them, not tearing them down. It is written with the understanding that we are all leaders in our own right within our circles of influence, irrespective of title or rank.

By learning and putting into use the principles and values covered in this book, you will improve your relationships, stand out from the crowd, and excel in your career.

The construction industry will face huge challenges in the coming years, but these same challenges will create opportunities for the leaders, who want to put in the work to solve the problems and their success will depend on their ability to connect and collaborate with others.

If you are reading this book, then you are one of these leaders. So, congratulations! Let's get going, there is a lot to do.

1

Human Connection – The Key to Influence

Relationships Will Determine Your Success

Relationships are fundamental to every interaction and transaction in your life and are the key to unlocking your potential for a successful future. This is true for both your construction career and your personal life. Regardless of situations you have encountered in the past or decisions you have made, it is never too late to start improving your relationships. It takes conscious effort, but you can do it starting right now. The fact that you are reading this book shows that you care enough to invest the time and effort to improve your life and I am honored that you have selected this book to be part of your journey.

Before two or more people are able to effectively operate with high technical function and collaboration (e.g. business, romantic, or otherwise), they must first make a fundamental human connection based on three key components: mutual respect, trust, and care. The word "mutual" means the feelings and attributes of the relationship are two-way and dynamic. Whether or not you are conscious of this, these connections start forming quickly after initially meeting someone. This is your first impression. How many times have you met someone and instantly felt connected? An inexplicable feeling that you were going to get along based on no prior interactions or information other than their name and occupation. The opposite is also true. Have you ever met someone for the first time and felt that something was off about them based on little or no information? Our brains have been hardwired to

The Human Side of Construction: How to Ensure a Successful, Sustainable, and Profitable Career as an AEC Professional, Second Edition. Angelo Suntres.

make snap judgments that were previously used to determine our ultimate survival but have since been dulled to just give us "feelings." While the topics of psychology, evolution, and the subconscious are important parts of building and maintaining healthy relationships, I am not an expert in any of those fields. As such, this book focuses on high-level principles to make the right connections and leverage them for success. However, you are encouraged to investigate these areas yourself for extra points!

I want to clarify that the intent of expressing and applying the skills in this book is to genuinely connect with people and provide mutual value for anyone and everyone that you cross paths with. It is not meant to use and manipulate people to get ahead. Similar principles can apply in either situation but you will not experience long-term happiness and success by manipulating or treating people like resources. It just makes you a bad person, so please don't do it.

Behind every personal achievement is a community of supportive people and connections that made it happen. There are few things in life that we can do alone and professional success is no exception. As you read through the following chapters, keep in mind that relationships can make or break your career and personal life. You can leave these decisions up to chance, or you can take an active role in prioritizing, planning and executing relationship management to increase your chances of success on site, in the boardroom, or at home.

While the principles and techniques described in this book can be applied universally, chances are you are reading this book because you are keen on accelerating your career in the construction industry. You may be asking yourself, "What do all these feelings and relationships have to do with a rough and tough industry like construction?" Well, that is a great question. Simply put, over the last decades the construction industry has evolved into a very transactional environment with little to no focus on enhancing "people" skills and the human experience. It is predominantly, "Here's your schedule, here's your budget, get it done." As a result, we have an industry of very technically strong people – brilliant architects, engineers, skilled trades people – people who have spent years honing their crafts based on concrete, quantifiable methods and principles but what academic course or professional training covers the topic of everyday life including interpersonal relationships, conflict management, and dealing with difficult people? All job descriptions and postings emphasize interpersonal skills as important for the role. Construction is rooted in working with people, lots of them, with many different backgrounds, training, and levels of experience. Yet, people skills are largely left to the individual to take on as part of their personal development.

What's funny is that if you ask anyone in the construction industry – project managers, estimators, superintendents, foremen, journeymen, and apprentices – what they spend at least half of their time doing every day, they will tell you that it is "dealing with people." Half the day is likely a conservative estimate. Depending on your role you may spend your entire day dealing with people and their problems. Why, then, are we not equipping ourselves and our teams with the tools needed to properly conduct our work in a psychologically safe manner? That is why topics like these are so important to learn. We will cover more on this in the next chapter.

The position we are in today poses a lot of opportunity for people like you who are willing to invest their time into building connections and improving the industry. This will be a huge distinguishing factor for you and will help you immensely to stand out and attract the right kind of attention. There has never been a more important time for the people of the construction industry to stop and think about how we treat each other and its direct impact on factors like productivity, profitability, and overall job/life satisfaction. It is time for a change!

This Will Improve Your Life

The main reason I wrote this book is to share what I have learned over the years to help you improve your current career situation and overall life. In this section, I will explain how learning and applying these principles will help you in your career. We will walk through some situations that you may identify with, and learn how leveraging relationships can benefit you and the people around you.

As an employee in a traditional company with a large organizational structure (more than four levels), you will have many types of relationships with many types of people – coworkers, peers, managers and friends – that you can nurture to maximize your potential for success. This is the most common situation and what you likely identify with.

Whether you are an entry-level employee, senior leadership or anywhere in between, there are three types of relationships that you need to be aware of: they are up, down and around you. These three types of relationships are equally important and serve different purposes for your success and can be illustrated as the three legs of a stool. Some relationships appear to provide more value as they connect you to higher power and authority but if any one of the relationships (the three legs) is compromised, the stool will topple and you will fall. Keep this concept in mind as we elaborate on the details below.

In large organizations, regardless of how big or fancy your title is, you will probably always have a boss. Even presidents are held accountable by boards of directors or shareholders. So, you must always be cognizant of "managing up." Establishing and nurturing upward relationships are the most critical in pursuing future roles and responsibilities because these are the ones that will have the most impact on your career. Because managers have the ability to make decisions regarding directly hiring/firing, promotions, and incentives, it is important to identify these relationships early and focus on strengthening them often as it takes time a repetition to form a meaningful connection and bond. In short, if your boss likes, respects, and trusts you, the probability of you advancing and succeeding increases.

While connecting with managers/leaders is critical and is the most effective way to set yourself up for future success, these are not the only relationships you need to reach your goals. Connection with your peers – or those of similar experience/status – will have the largest impact on your day-to-day enjoyment and job satisfaction because you spend most of your day working with and around them. These relationships are extremely important. Gaining trust and respect in your coworkers will not only make for a happier, more enjoyable work experience but it will also increase productivity and effectiveness in your team. All of these results will eventually make their way back to your boss as well, thereby strengthening that relationship. The key takeaway here is that how you treat your coworkers will have a direct impact on your future success. This is also where you may have the biggest challenge and run into the most adversity as there are a lot of different personalities, especially on large teams. Some team members will even have different or questionable agendas; we will learn more about these situations in a later chapter.

The third and final relationship type in a larger organization is your connection with junior employees. These could be direct reports, new hires, or just those with less experience with whom you have some amount of responsibility or mentorship. While these relationships will have the smallest bearing on your personal career compared to the other two types, I genuinely believe that this is the most critical relationship from a leadership standpoint. As leaders, we have the responsibility to ensure the people in our fold are given the tools and environment to reach their full potential, to the extent that they are willing to receive the value you can provide to them. Focusing on this relationship also shows integrity – doing the right thing even when your boss is not looking – a quality that many leaders possess and look for in others.

If any of the relationships mentioned above are compromised or neglected, people will detect misalignment in your character and this will limit how far you go in your career. So, be mindful of all relationships, no matter how insignificant they may seem.

Another situation to consider is if you are a contractor or solo entrepreneur – you are the boss, the one and only. Well, guess what? Your customer base and anyone that you need to hire to scale your business or achieve work/life balance will rely on your ability to interact with people. While you may have little or no relationships required within your business, the business itself will rely on how you build and maintain your relationships with customers, suppliers, and subcontractors, so the same rules apply!

In summary, people conduct business with people they know, like, and respect. This goes for coworkers who enjoy working with you, external customers who choose to give you their money, and even your management team who truly hold the key to your future.

How you treat people is the biggest reflection of you as a person and will ultimately determine your success.

Now that we have covered how building and maintaining healthy relationships and focusing on how the human elements of construction can benefit you personally, let us talk about how it will help the industry. Imagine a world where project designs are always 100% complete, estimates are always perfect, all procurement comes in below target, and construction is always completed on time and within budget. Okay, that is never going to happen, but allow me to explain how improving our relationships can help us get closer to this utopian dream.

Let us face it: the construction industry is struggling for a lot of different reasons that we will not cover in this book. Possibly, the largest challenge right now is that there seems to be a lack of trust at all levels in the industry. Owners don't always trust contractors, even in some collaborative delivery models (it is ironic but true). Gone are the days of making deals based on handshakes and owners having blind faith that contractors have others' interests in mind, and not just their own profits. This has been spoiled by a few bad apples who may have taken advantage of certain situations like change orders and contract loopholes. There also seems to be a lack of trust between contractors and subcontractors resulting in constant fights about money and scope. The bottom line is that there are challenges and conflicts that we face daily with little to no tools or resources on how to deal with them.

Faced with difficult situations, many people are elevated to a state of "fight or flight," reacting as if they have been threatened or attacked, and most of us in the industry choose the former. If you reflect on your most recent heated meeting or conversation (I hope you have not had many, but they are common), what was the cause? Usually, a comment is taken personally and triggers an elevated emotional state. This adds a layer of complexity to the original problem, which likely could have been solved easily without the added drama. In my experience, problems are easy to solve, but people make them difficult because everyone carries around their own agenda, experience, and baggage, which adds "stuff" to deal with. Most of this is subconscious but if you take time to understand and appreciate where people are coming from before reacting, you can avoid wasted time, effort, and embarrassment.

Diverse types of people are involved at every step of the project lifecycle from project planning, conceptual design, procurement, and detailed design all the way to execution, commissioning, and handover. All types of people, old and young, technical and commercial, leaders and followers are required to coexist and collaborate. The success of every project from start to finish is contingent on how well the people work, communicate, plan, and execute as a team. On a larger scale, the success of a group of projects over time across different companies and clients advances the collective industry. Therefore, it is reasonable to believe that if we can improve how everyone treats each other – their relationships – day in and day out, across all companies, clients, and projects, the industry will naturally improve as well.

It may be a stretch but I believe it is possible, and it starts with you.

How It Worked for Me

My career in construction began in 2007 after graduating from the University of Western Ontario with an undergraduate degree in mechanical engineering. During the later years of my studies, I specialized in HVAC design and after graduating was determined to be the best mechanical engineer the world had ever seen. When it came to finding a job though, the economy was just coming out of a recession and opportunities were scarce. I ended up in a situation where I just needed to start making some money, so I started applying for positions outside of the engineering consultant sphere. Call it coincidence, luck or whatever you want to but I ended up landing my first full-time role at a mechanical and electrical construction company. I originally took the job out of desperation but, looking back now, I could not picture myself doing anything different today and am eternally grateful to whatever force placed me in the industry. My first role included a mix of estimating design and project coordination, which was a fantastic opportunity to transition into the workplace, apply the knowledge that I had learned in my studies, and see the real-world impact while collecting a paycheck! It also gave me great exposure to the worlds of estimating and project management and allowed me to develop a basic understanding and appreciation for both ends of the project lifecycle.

The first 10 years of my career were focused on honing the technical side of my craft. I spent a lot of time learning about the basics of construction including logistics of sites, how to apply the science that I learned while studying engineering, and how to manage budgets and schedules. I quickly learned that construction is a highly technical field and there is a lot to learn. It was at this point I also realized that there was an enormous amount of coordination involved, which is where my desire and love for leadership and communication were born. I found that there were many gaps in how things were communicated and this really got me thinking. I started to notice that a lot of issues we encountered – on site, in the trailer, or in the boardroom – were a result of communication challenges. I also noticed that a lot of people were mistreated by weak leaders in the industry and wondered why.

While transitioning into my first management role as Chief Estimator, I really started to focus on the interpersonal side of the business, taking interest in what motivates people to achieve their best and how they work optimally in a team setting. I was later promoted to Manager of

Estimating and Pursuits where I oversaw a total of over $700 million worth of mechanical and electrical pursuits. This was invaluable experience where I was able to nurture existing relationships with clients, suppliers, and coworkers as well as build some new ones in the process. Utilizing the principles shared in this book helped me succeed. In my current role. I work for a prominent general contractor where we design and build some of the biggest, most complex projects in Canada.

From a young age, helping people was a passion of mine; it brought me immense joy. In this book, I have condensed all the tips, skills, and techniques as well as successes and failures that I have experienced in my years in construction. It covers the good stuff that worked and the bad stuff that did not in hopes that it will save you time, effort, and energy in achieving your goals. I have personally experienced huge successes in both my personal and professional life and I have also seen the dark side of the human experience. I have been hired, promoted, laid off, and fired. So, wherever you are in your journey, I understand and acknowledge you and I know it can and will get better. I have experienced the challenges of trying to find work early in a career, managing external stresses like family and other relationships, pivoting mid-career through a pandemic, and transitioning into management roles. I do not share this information to brag, or to get pity, or to seem like I know more than anyone else. I am simply here to share my experiences in hopes that it can improve your life the way it has improved mine.

I am confident this book will bring happiness to everyone who reads it, but if by writing it, I manage to help even one person, I consider it a success.

Of course, everyone is free to make their own choices and have their own opinions. The information I share in this book will not harm you in any way. If applied correctly, it will provide huge personal, professional, and financial success. At the very least, it will make you think long and hard about what experiences you have had in the past, how you can use them to attain experiences you want in the future, and the skills and behaviors required to accomplish your goals.

In short, you do not have to listen to me but if you do your life, your career, and our industry will be a better place! Plus, if you are reading this, you have already purchased the book. So, you might as well keep going.

How It Can Change the World

The principles of human connection, effective communication, and relationship building are not confined to the construction industry. They are universal, transcending industries, and cultures, and have the potential to bring about significant change in the world regardless of the sector you work in, whether it is architecture, engineering, or ownership, and beyond. In a world increasingly driven (and divided) by technology and automation, the human element often gets overlooked. Yet, it is this very element that forms the foundation of our connections to others and the underlying fabric of society. The principles discussed in this book, if applied beyond the construction industry, will lead to a more connected, empathetic, and understanding world at a time when everyone needs it.

Effective communication can bridge the gap between different cultures, fostering a sense of unity and mutual respect. These are the key elements to team building regardless of your industry. It can also help resolve conflicts, paving the way for peaceful collaboration, not defensive and divisive arguments as is the default mechanism to deal with conflict in a lot of workplaces usually driven by ego. In the corporate world, it can lead to more inclusive workplaces, boosting employee morale and productivity, resulting in happier people and better team performance.

Relationship building, on the other hand, can lead to stronger communities. It can foster a sense of belonging, reducing feelings of isolation, which, especially as we are still feeling the effects of the COVID-19 pandemic, are critical to achieving harmony and balance in our personal lives. In business, it can lead to stronger partnerships, driving innovation and growth and facilitating better communication as outlined above.

Moreover, these principles can play a crucial role in addressing one of the most pressing and widely discussed issues of our time, social inequality. At its core, the Human Side of Construction is about mutual respect and inclusivity regardless of background, age, or place of origin. It is about treating others as you would like to be treated, a simple notion that has seemed to be forgotten by many in the anonymity of the internet and faceless names of large populations. By fostering a culture of collaboration and open communication, we can come together as a global community to find solutions and make the world a better place.

In essence, the principles of human connection, communication, and relationship building have the power to transform not just the construction industry, but the world at large. They can help us build not just physical structures but also bridges of understanding and cooperation in all sectors and all corners of the globe.

As we navigate through the complexities of the 21st century, let us remember that our success, both as individuals and as a society, will depend on our ability to connect, communicate, and collaborate. The time to embrace these principles is now. The world is waiting.

2

Your Network is Your Net Worth

Develop Your Support System or Perish

> If you want to go fast, go alone. If you want to go far, go together.
> African Proverb

When first starting your career, you may be one of the lucky few people, who have a friend or family member, or who can provide a direct reference, but most people get their first job strictly based on what they know since they have no connections in the industry. You apply for a position, demonstrate that you meet the basic requirements on your resume or CV, and with a little luck, you get selected for a meeting. From this point on, starting with the interview, relationships will be the most important determining factor in the level of success you will achieve in your career. It may sound ominous but as you will find in the following pages, it is quite simple when you know what to do and how to do it.

Before you can start to establish strategic connections with people and build your support systems for success, you will need to have a clear vision of what your desired career path is including any other achievements you want to accomplish. Setting goals and establishing a plan to achieve them is critical to start you off on the right course. If you require a bit more reflection on this area, the appendices provide a step-by-step process, which will help you determine what works for you based on three key questions. Feel free to skip to that section now or continue reading if you would like to do it later.

The Human Side of Construction: How to Ensure a Successful, Sustainable, and Profitable Career as an AEC Professional, Second Edition. Angelo Suntres.

Presuming that you know where you want to go, it is time to discuss how you are going to get there, and most importantly, who is going to help you. In a complex, technical industry like construction, it is impossible for any one person to hold the knowledge to answer every question and solve every problem. Do not strive to be this person; it is impossible, and you will fail at some point. Rather, be the one who knows where to go to get the answer or at least to start the journey to the right person. Your success in the industry will depend on the connections you make and keep throughout your career. These will benefit you in many ways, namely with providing technical assistance and advancement opportunities. On the topic of technical assistance, if you build your network strategically and correctly, you are only ever two to three phone calls away from answering any question or solving any problem you might encounter. Regarding advancement opportunities, people must know and like you to work with you. So, again, correctly building your support systems will help with career progression.

I believe that every person on earth can somehow bring value to our lives and ours to theirs in return. This applies even to people you do not like or get along with; they just take a little longer. The theme here is not sucking up to your boss/senior management or using people to reach your goals but rather building strong and mutually beneficial relationships, strategically. Obviously, you need a good rapport with your manager and anyone else who is ultimately responsible for making decisions regarding promotions, salary increases, etc., but there are other key supports that you will need to succeed, which are covered later in this section.

Generally, you will find that some bonds are easier to form than others – some are even difficult – but it all comes down to treating people with respect and expecting the same in return. Any connections you make will not only help further your career but will also help the day-to-day enjoyment of your job, which, in turn, will improve your life. This does not mean that you must be best friends with everybody but be kind and respectful to everyone around you and develop a reputation for being a connector. At first, this may not seem natural to you but once you start opening up to people and having them do the same in return, you will gain momentum and it will become easier. We will discuss ways to connect with others in a later section.

When considering which relationships to prioritize, you need to have a strong understanding of the organizational structure of your company and to determine who the decision maker is at different levels. Who you will need to connect with at each step of the way will depend on your career goals. For example, if you are a Project Coordinator and your immediate goal is being promoted to an Assistant Project Manager, it is important to prioritize a connection with your direct supervisor, usually the Project Manager, as they will have a large say or total control in helping you make this transition. If, at the same time, you have ambitions of moving into larger roles or responsibilities in future steps of your plan, it is also important to plant seeds with higher levels of the organization; however, this can be a precarious situation as you do not want to overstep any boundaries by going over your immediate supervisor's head and causing upset. Generally, I find it best to communicate openly and clearly with everyone, assuming your intentions are good. Using the example above, it is perfectly fine to discuss with your immediate supervisor (Project Manager) the topic of one day moving up to their level while, at the same time, speaking with them or other senior leaders (Construction Manager/VP) about your intentions to be promoted to even higher levels in the future. Provided that these goals are accompanied with realistic timelines and appropriate steps, they will show ambition and work in your favor.

Few things in life are worth doing alone. People are meant to connect with others; it is in our nature and how humans have been living for thousands of years. Lately, there has been a lot of talks around building and participating in a sense of community for various aspects of life – social, hobbies, work, etc. While this is especially important following periods of isolation from others for months during the pandemic, this is not a new idea. The old saying, "It takes a village to raise a child" is still true to this day and can be applied to construction: it takes a team to build a building. Finally, taking it one step further to align with the theme of this book, it applies to your professional life: it takes a network to advance a career.

To build a successful and long-lasting career, you will need support from all angles, achieved through your network and composed of four types of relationships outlined below.

1) Management (Direct Supervisors/Managers, Senior Management)
   - This is the smallest but mightiest connection group. There is usually a small number of managers compared to the overall company population. Also, you will likely have less frequent interactions with the management team, depending on your position.

Regardless, this is the most critical connection to make as it will have the biggest impact on you (good or bad!)

- Remember that it is ok to connect with people beyond your immediate manager but it should be in an indirect way to avoid any potential conflict. An example of an indirect connection to more senior management is if you both volunteer for the same groups or play together on the company softball team, etc. Also keep in mind that you do not need to be best friends, you just need to be visible and memorable (in a positive way)
- This group will have the most leverage on your career as they have the capability to make decisions that directly impact your career

2) Peers (coworkers, team members, different department members with similar seniority level as you)
   - This will be the most available type of relationship because there are lots of people that you know, or come across every day, that fall into this category
   - These types of relationships will have a big impact on your career for two important reasons:
     - they help improve the day-to-day enjoyment and productivity of your current role (personal)
     - when considering people for promotions, the leadership team usually assesses how people act with their peers, how they are respected and how they work in the team (professional)
   - Plus, it is always just nice to be a good person!
3) Junior (direct reports, new grads, different department members with less seniority than you)
   - While this type of relationship will likely have the least impact on your career, I personally believe that it is the most important of them all. The way you treat someone who brings little to no direct value into your life is the greatest true reflection of your true self. Countless times I have seen junior team members or apprentices hazed, disrespected, and belittled. There is no place in any work culture for that; so if you see it, say something. Okay sorry, I had to get that out of my system...
   - If you have the responsibility of managing a team, this is a critical area to focus on for two reasons:
     - If you take care of your team, they will function well, and this will reflect positively on you

   - If your team is high performing and taken care of, chances are they will be happy and they will become your biggest fan/supporter. This information will also make its way to the right people and help your career

4) Family/Friends
   - So far in this book, we have focused a lot on building relationships inside of your work ecosystem; however, it is just as important to have relationships outside of work
   - While friends and family will not directly help you advance your career – assuming they are not in high-power positions and can "pull some strings" – it is imperative to have an outlet outside of work that you can rely on. There will be times when you will encounter people or situations at work that you carry home with you. If you do not properly deal with these issues, it will negatively impact your life both in and out of the office/site
   - By ensuring that you have the right people that will support you outside of your career, it will improve your overall life as well

How to Engage with People Effectively

Identifying, contacting, and engaging people is one of the biggest challenges that many people face in their careers, especially when it comes to mentorship. In most organizations, there is little to no training given on how to network and connect with people so it is left up to the individual to do their own research. Some are born with a natural charm or charisma that naturally draws people in; however, most have to work to achieve this level of personality and even those that are born with it have to work at other supporting qualities like communication. Furthermore, if you are a charismatic, likeable person, but you do not express qualities like integrity, accountability, or basic knowledge to complete your job, your success will be limited. This presents an opportunity for you to differentiate yourself from others – even if you are only completing the basic requirements of your job – by becoming a master at communicating and connecting with people. Even if these skills seem unnatural to you or you think that they are out of your character, as long as you have the willingness to learn and dedication to practice, you can develop and leverage them to help advance your career.

So, how do you find, contact, and engage the right people? This might be easier than you think, and does not require putting in lengthy additional hours at conferences, seminars, and social events. The key is to make an impression on people consistently in both frequency and quality of interactions. While a single outstanding task or presentation will get attention, it is familiarity over time that will bring the most success. A compounding set of frequent and positive interactions will create a positive association or reputation.

The most important thing to keep in mind is that the objective is putting yourself in the right situations with the right tools and the right people. When fishing, you cannot just walk down the street or go for a swim and expect to catch a fish; you need to have the proper equipment and then go to the right place. The more prepared and deliberate you are, the higher your chances are of getting what you want. Structuring your relationships for career success is no different; you need to learn and apply the tools and techniques covered in this book and put yourself in the right situations.

Let us start with the first example of connection we mentioned above: the management/leadership level. If you are a Project Manager or Coordinator working on a specific site, you likely have little day-to-day contact with people outside of the project. The Construction Manager or Vice President might come down to site once in a while but usually only if there is a major problem, at which point they will probably not be in the best mood to connect! Getting involved in activities outside of your main role is a great way to connect with others. Examples of this could include any volunteer efforts that your company supports either internally (social committee, environmental committee, etc.) or externally (Habitat for Humanity, etc.). These are great opportunities to meet people and show that you care and have interests beyond just doing your job and collecting a paycheque. Taking part in activities like this does not have to extend beyond your normal hours either; many initiatives like this occur during working times so you do not have to sacrifice family or personal time if that is not an option.

The same rules apply when trying to find a mentor, which is covered in more detail in the next section. Be open to meeting people and naturally curious as to what they can offer you, and what you can offer them in return. Even mentorships should be a two-way street. Mentors do not need to have the exact experience that you hope to gain. So, do not get too stuck on types of people or careers. In my experience, this type of relationship is stronger if your personalities match rather than having similar job titles or fields. Mentors can also serve as hubs that can direct you to others in their network if a specific need is required that they cannot address.

When approaching people for mentorship, it is important to clarify expectations at the outset to avoid disappointment or frustration on either end. Also, do not expect people to offer up mentoring you as many see this as a serious commitment. Start with a professional connection. See if you develop a personal bond and also see where it takes you. You do not have to make a formal proposal like a wedding engagement!

The best way to meet and connect with others is through direct and frequent interactions with common values such as the examples in the previous section. However, there are some additional ways to connect that may still be effective, though to a lesser extent.

Direct messages can be a terrific way to make an initial introduction either through email, LinkedIn, or other messaging platforms. If you decide to go this route, it is most effective to keep it brief. Everyone receives a vast number of emails and messages; so, you should stick to the point and invite further conversation to get into more detail. You could explain who you are, what you do, and why you would like to connect. Focus on what value it would bring the other person; so, it does not look like you are just trying to get face time for your own advancement. Here is an example of a message that could get a conversation started:

> Subject: Interested in Connecting
>
> Hello Mr. Smith,
>
> My name is *[your name]* and I am a *[your title]* at *[your company]*. I am very interested in *[something they are working on]* and feel that I could bring value by *[how you would bring value]*. I would love to connect and discuss more when you have the time. Please let me know when you would be available.
>
> Thank you,

The less the person knows about you, the shorter the message should be. They likely do not have time to read paragraphs of what you have accomplished or hope to achieve, and they are probably not interested at this point. Leave this information for the follow-up. When you get a response, follow up as soon as you can to take advantage of their attention and focus on setting up the next encounter whether that is a phone/video call, in-person meetup, etc. Because this is a cold call, you may not get a response. If this is the case, do not get disappointed or upset; the other person is either too busy to respond at this time or the connection just was not meant to be. If you do not get a response, it is ok to follow up a week or so later as a reminder if it is a more convenient time for them. If there is still no answer after the second attempt, it is time to move on as you do not want to come across as a bother.

Social events can be another great way to get in touch with people. These could be internal company events like a holiday party or an industry conference where there are many different types of people from all levels of organizations. While the main intent of these functions is to bring people together, in my experience, it can be challenging just due to the overwhelming volume of people in attendance. You may get to spend a few minutes with someone but it is difficult to have any quality time, especially with management as there is usually a line of people trying to get their attention. To stand out from the crowd, do not waste time chatting about the weather or how nice the appetizers are. Have an agenda, leave a quick lasting impression, and offer to chat more about it later or ask if it is okay to contact them the next day to set up a future conversation. Nothing beats human-to-human interaction!

Do You Really Need a Mentor?

A mentor is defined as an experienced or trusted advisor. Essentially, this is someone who embodies two key qualities:

1) They have had relevant experience and can provide some input, advice, or guidance to help you with specific problems or larger planning ideas, etc. Usually, a mentor is someone who has been in a similar role as you but this is not necessarily a requirement. Anyone who has been in the industry for a few years can relate to and give guidance on many different situations; however, if you are looking for step-by-step solutions to specific problems, you will likely require more a coach than a mentor. A mentor is someone who provides high-level advice and while they likely can provide detailed

instruction on certain tasks, this probably is not the most effective use of their time.

2) As we have discussed before, trust is a fundamental aspect of every relationship and this is extremely important in establishing a mentor. Ideally, a mentor is not a direct supervisor and is an impartial third party to your career. This avoids conflicts of interest for everyone and will provide you with a safe outlet to discuss internal/team challenges that would not be appropriate with your manager. Also, if you are going to invest time and energy with someone on top of your career and personal life, you will want to make sure that they are trustworthy and have your best interests in mind. Otherwise, you may be wasting your time.

Mentorship is an important topic, but many people struggle with it in their careers. Some organizations have formal mentorship programs, but most do not. Largely, the responsibility of finding a mentor will fall on each individual. In my experience, my best mentors have come through mutual contacts and when I was not looking. This is another example of how being open to networking and connecting with different people is important. You never know how you are going to meet a mentor or how they are going to impact your life.

Most people I speak with say they have not had a mentor and that they have missed out because of it or that they have not experienced as much success as they could have. Having a mentor is not a prerequisite to success. So, do not feel bad if you have not had the opportunity to connect with someone on this level. In fact, if you think about your closest work relationships, chances are that you probably do have at least one or two people in your life, who have acted, at one point or another, as your mentor. It does not have to be "official" to be beneficial.

For the more experienced readers, consider yourself earlier on in your career. How much time and effort could you have saved if you had someone guiding you along the way? It is an excellent way to give back to others and the industry and does not necessarily take up a lot of time. You do not even have to label it as a mentorship; just spend time with people who need a bit of help.

Now that we have covered how to identify the types of people and relationships that you will need in your support system, in the next section, we are going to discuss how to engage with them to get the best results.

How to Diversify for Sustainable Success

In the dynamic world of construction, diversification is not just a strategy; it is a necessity for sustainable career success. As the industry evolves, so too must our approach to building and nurturing relationships. The construction industry is a melting pot of diverse skills, backgrounds, and perspectives and if you learn how to leverage all of these components to your advantage, you will have an edge on your colleagues and experience great success. This diversity is a strength that can be harnessed to foster innovation, drive performance, and build resilience. However, to tap into this potential, we must first understand and appreciate the value of diversity and how it can impact your network and career.

Diversity in your network means having connections with people from different disciplines within the industry, and outside of it, at various levels of experience. It is impossible to know everything about one single trade in the industry, let alone all facets of construction. What you can do, however, is know who to look to, to attain the information you need. This is the heart of effective networking; building a web of individuals that serve you and whom you serve in return. It is about broadening your perspective by engaging with people who think differently, challenging your ideas, offering fresh insights - complementing your strengths and supplementing your weaknesses. This diversity fuels creativity, fosters innovation, and can lead to more robust solutions to the complex challenges we face in the construction industry.

Building a diverse network requires intentional effort. It involves some strategic thinking, stepping out of your comfort zone, seeking out new connections, and nurturing these relationships with genuine interest and respect. It is about being open to learning from others, sharing your own knowledge, and finding common ground amidst differences. The key is creating a dynamic, two-way relationship where both parties benefit. One-sided arrangements rarely stand the test of time as people catch on quickly and do not like being used for information or progress.

Authentic relationship building is at the heart of this process. It is about more than just exchanging business cards or adding contacts on LinkedIn. Authentic relationships are built on mutual respect, shared values, continued contact, and a genuine interest in each other's success. They require consistent communication, active listening, and a willingness to invest time and effort. This may sound like a lot of work, and not many people are naturally versed in making these connections, but the rewards can be tremendous.

In the construction industry, these relationships can be the difference between a project's success or failure. They can lead to collaborations that solve complex problems, innovations that push the boundaries of what is possible, and a work environment that attracts and retains top talent. The same is true for your career. If you build a reputation for building authentic, mutually beneficial relationships, people will want to work with you and for you; but the opposite can also be true if you operate with the intent to use people to get ahead.

As we navigate the challenges and opportunities of the construction industry, diversifying our networks and building authentic relationships will be key to our sustainable career success. It will enable us to adapt to change, seize new opportunities, and build a future that is not only professionally successful but also personally enriching and fulfilling. Remember, the strength of your network is not measured by its size, but by the diversity of its members and the quality of the relationships you maintain.

3

Basic Human Connection Skills

Trust – The Foundation

Trust can take years to form, seconds to break, and forever to rebuild.

In any relationship, business or personal, trust is the most important factor. It is essential to building a long-lasting, successful career. Without trust, relationships become volatile, unstable, unhealthy, and counter-productive. What makes this topic complicated is that, in many cases, it is not as simple as starting from zero with new connections. Many people, especially in the construction industry, have had negative experiences in the past that resulted in disappointment, anger, or pain. This can make trust difficult to give and, as a result, to earn. Trusting others means you are vulnerable to being hurt or taken advantage of and if you have experienced this before, you will probably try to avoid it in the future. While this book does not intend to cover the topic of human psychology, having a basic understanding of these concepts will provide an advantage in considering how to connect with people.

Regardless of your experience or that of the people you try to connect with, it is possible to establish trust. It may take different amounts of time and effort depending on your past interactions or reputation, and the level of trust required to accommodate the relationship. For example, negotiating a contract for millions of dollars will take more trust (time and effort) than purchasing a $10 eBook. Trust is built up over time by showing the other person that your behavior is consistent and in the best

The Human Side of Construction: How to Ensure a Successful, Sustainable, and Profitable Career as an AEC Professional, Second Edition. Angelo Suntres.

interest of the relationship (reciprocal, not singular). To illustrate this concept, consider your current role and picture that you regularly and satisfactorily complete tasks assigned by your manager in a timely manner. This is the most basic way to build trust and rapport as your manager knows that they can count on you to complete your work and fulfill your role in the organization. In this example, establishing trust could bring an opportunity to take on more responsibility or advance your career. This may be simplistic, but you would be surprised how far you can get just by completing your work pleasantly, properly, and on time. Many people struggle to do all three.

Another crucial factor of trust in a relationship is the handling of sensitive information. If someone opens up to you in confidence about a personal or work-related matter, the best way to break that trust is to share it with somebody else. Likewise, if you are exposed to information such as trade secrets or government projects as part of your employment, you are awarded a certain level of trust to protect this data. Breach of this trust will not only have personal or career impacts but can also come with legal action as well. Whether it was explicitly stated as sensitive information or not, you need to exercise good judgment to make sure you do not damage the relationship (as tempting as it might be to share some juicy gossip ... hey, it is human nature). Trust can be viewed as a two-way street and reciprocal, meaning that whatever you give in a relationship, you should expect in return. Generally, in life, this can be summed up with the Golden Rule: treat others as you would like to be treated.

Before you can conduct business, collaborate, or solve technical issues (the basic qualities of construction), you need to establish a strong foundation with other people involved; it requires a human-to-human connection. Without this solid relationship base, you will struggle with the more complex matters of business and ultimately with success in your career. For the skeptics, who believe that this is all too warm and fuzzy and believe that we should all just keep our heads down and get our work done, I will refer back to how human civilization has conducted itself for the last few millennia. People have always relied on the support of others through communities and groups for all facets of life – raising families, developing businesses, and, of course, building things. There are no positions in the construction industry in which you will not interact with other people. So, it is critical to acknowledge and develop interpersonal skills including building and maintaining effective healthy relationships.

Trust is the principal factor in human connection and can have the strongest effect on your relationships with people. Consider those that you have connected with in your life so far. Surely, there are many positive experiences, but you can likely think of one or two people who, for one reason or another, left you with a bad impression. Have you ever stopped to think about why? It likely comes down to an instance or instances where they mistreated you, held different opinions or viewpoints, or just rubbed you the wrong way. In any case, that particular instance resulted in a feeling that you simply do not trust the person, especially with something as important as your time, energy, or financial resources. At the core of every human, we all want to feel valued, respected, and loved and by accepting any of these from another person, we have to trust that they are sincere in their actions and feelings and that they have our best interests in mind.

Here are two important factors to consider in your relationship and how they impact the dynamics of trust:

1) Your ability to trust others. Events may have happened in your past that affect your ability to trust others, especially in specific situations. This is perfectly normal, especially if it causes you pain, suffering, or trauma. Still, know that despite these bad experiences you may have had, there are a lot of good people out there who mean well and are worth connecting with. You might require some extra steps to get to a point where you can start trusting others but there are resources available to help you get there. There is nothing wrong with seeking help from others in the form of therapy or just talking to close family or friends that you know will support you.
2) Others' abilities to trust you. Consider your past for a moment. Have there been instances where you have broken other people's trust? Have you been unreliable in completing tasks at work? Many people, myself included, who take the time to consider this might not even realize that they have been hurting themselves. Like point number one above, it is never too late to change and it starts with acknowledging that there may be an issue. Just because you have had bad experiences does not mean you are doomed to be unsuccessful or repeat the mistakes you made in the past. You have the power to change.

Simply put, there is no relationship without trust. It is an elemental building block forming the foundation of your connection with others and its importance cannot be understated.

Communication – The Glue

Communication skills are a must in any type of relationship. Like other skills we cover in the book, there is very little formal training on how to use effective communication, especially in construction, and many people assume that they are doing it correctly. We could all use some help in this area as you can never be too good at communicating and, if you learn to do it well, it will have the most impact on your career since it is so severely lacking in construction. Whether you think you have superior communication skills or are unsure about your current level of sophistication, the best test is to ask those who you spend the most time with – family, friends, spouse, etc. Personally, I have learned the best communication lessons from navigating family logistics with my wife and children! The same principle applies in any relationship though; be clear and concise in your communication whether verbal or written. Whether you are stating your intentions, requesting permission from your manager or providing an update of sorts, communicate clearly and be direct to avoid confusion or misunderstanding.

The critical points here are:

1) Delivering the correct information
2) Through the correct type of media
3) To the correct people

It seems cumbersome but before sending out an email or arranging a meeting, consider these three questions to determine the most effective way to communicate. Selection of the wrong delivery method or inclusion of "fluff" will result in a loss of impact, and maybe even understanding, of your message. It initially takes some time, but this will become a habit and will ultimately save you and your coworkers time.

Knowing which method of communication is best will differ depending on the situation you are in. If you are providing a quick update that requires very little context or summarizing an event or conversation, then written communication in the form of an e-mail or text message will likely suffice. Make sure you keep things concise and to the point. Unless you were asked to elaborate on the details of what, when, where, why, and how, you do not need to include that information. Follow directions and address questions clearly to avoid confusion and frustration.

Verbal communication is preferred if there is a certain situation or update in which you are required to provide more than one or two quick details or elaborate on context. An example of this is if you are addressing a power outage on a certain level of a construction site. Depending on the stage of construction, usually people want to know the root cause of the outage, what was done, and how you are going to prevent it from happening again. This is best done in a phone call or meeting with the appropriate stakeholders since you will need to deliver detailed points to different people who will likely have questions. Similarly, any situation that requires collaboration or input from others should be done verbally and preferably in person – nothing promotes creativity, brainstorming, and connectedness like being physically present with someone and feeding off each other's energy. Usually, these types of meetings or conversations are followed up with an e-mail or formal set of minutes to capture the discussion points. Two-way conversations need to be done verbally and should avoid email chains that can quickly spiral out of control.

Verbal communication is the most direct way to clearly deliver your message. Keep in mind there is so much more to communication than words (i.e. nonverbal cues) and, as humans, we all have lots going on in our lives. Perhaps they are having a challenging day, or they are overwhelmed with other tasks, or maybe their dog just died ... you just do not know unless you are there with them at the moment.

It is better to be honest than to be right. Many people are so worried about saying the right thing that they are afraid to be honest, especially in dealing with matters involving their manager or even coworkers. Unfortunately, in many organizations, workers are not encouraged to speak their minds. Social pressures and toxic management force them to conform. If you establish healthy relationships with others, you should feel free to express your thoughts about most topics without fear of punishment and expect the same from others in return. The fact of the matter is that sometimes the truth hurts, and while procrastinating or hiding, it may alleviate pain in the short term. Eventually, it comes out and the longer and harder you have hidden it, the worse the impact will be.

My favorite saying on this topic is, "Go awkward early." If you know something is not right, even though it was a mistake you or the team made, it is in your best interest to let the appropriate people know immediately. Even if you believe you have a plan to recover or solve the problem ... communicate, communicate, communicate. Yes, you will have to face a difficult conversation, but it will not get any less difficult as time goes on. Often the actual outcome of this type of conversation is not as bad as you make them out to be in your mind. I know in my experience, stress and anxiety have a funny way of connecting the dots in your head to make you think the worst is going to happen. If you are dealing with a reasonable leadership group, they will appreciate the honesty and if you are not dealing with reasonable people, it is a good time to find a group that is.

Now I am not encouraging anyone to be ignorant, pick everything apart, and tell everybody what is on your mind. It does involve some tact and gauging seriousness from minor items and who should be spoken with for each. It is important to use your judgment on what issues are worth pursuing and what is better off left alone. Also, if the person you are trying to communicate with makes it clear that they are not interested in receiving feedback, consider backing off so you do not come across as overbearing. If it is a serious problem that is being neglected, you may need to escalate the issue and speak with different people. Your duty lies in upholding moral values and doing the right thing for your company, not pleasing people.

A perfect example of this in construction is updating the progress of a project. At some point in their careers, Project Managers will realize that things are not proceeding according to schedule and/or budget. Some are inclined to not report "minor" delays or budget overruns with hopes that they will be made up in the future. In my experience, it is always better to report these issues as soon as they are detected. You could notify your manager that there is an issue and you would like to schedule some time to discuss it in detail. Prior to that meeting, you can meet with the team to review what caused the delay/overage and mitigation or contingency plans. This way when you meet with your leadership team, you are not just presenting a problem but also identifying the cause (to avoid this happening in the future) and plans to mitigate (to recover). This is a perfect example of how to demonstrate honesty, integrity, foresight, and care for your project.

Empathy – The Ultimate Connector

Simply put, empathy is the ability to understand and share the feelings of another. This does not come naturally to most people but it is critical if you want to build effective relationships. Everyone, regardless of age, gender, background, etc., brings with them their own experiences, opinions, biases, and baggage. Yes, even you! Empathy and care go hand in hand and people are more than just a cog in the machine. Yes, we all have our jobs to do and those functions, when completed together, complete the overall business objectives of an organization but there are people and personalities behind every role. For many, it is difficult to compartmentalize your personal life when you enter the office and since we are our own entities, what you do for a living does not necessarily define who you are as a person. Acknowledging this and connecting with coworkers on a personal level will do wonders for your relationships as it shows that you care for and value them and that they are not simply a resource to get a job done. This will instantly boost trust in your relationships, and the best part is that it requires very little time or effort.

How many times have you logged onto a call or entered a meeting room early and sat in awkward silence with one or more people before the meeting started? Or crossed paths with someone in the kitchen while you waited to heat up your lunch? It is in these unavoidable moments that a little bit of extra effort goes a long way in connecting with people. These are the perfect opportunities to ask people simple questions about what they did over the weekend and upcoming holidays or how they are dealing with a recent change at work or at home. In these conversations, always look for commonalities that you can bond over. For example, if someone mentions that over the weekend, they took their children to an event, and you also have children, now is a good time to share that information to bond over. Similarly, if they say that they went to a hockey game and you are not that interested in hockey but enjoy basketball, you can still bond over your love of sports. All these seemingly little ties will compound to strengthen your relationship over time.

Another example of the importance of empathy is when you are in a situation with another party where you are trying to reach alignment on an issue – it could be salary negotiation, solving interferences onsite, etc. – you must be cognizant of the fact that each side of the argument is pursuing their own best interests. Trying to understand where the other person or party is coming from will help you strategize your approach to reach a decision or conclusion quicker and with less stress for everyone. By taking some time to review why someone is making a certain argument, you may uncover more commonalities with your ideal results and even come up with alternate solutions. An example of this could be if you are negotiating a salary increase with your manager. First, it is worth noting that if you have a strong relationship with your manager, these types of conversations do not have to be awkward or uncomfortable. Let us say you would like a salary increase because of the increased cost of commuting since your initial employment. Understanding that your manager may have limitations in what increases they can grant and that they also have a boss to report to, maybe you can seek other ways to get your result. Working from home part-time or perhaps an increased travel allowance rather than base salary are two ideas that could be considered a win-win.

Expressing empathy to others is basically just caring about them as a person. This might sound basic and simple, but it may not be as easy as you think. In the construction industry, we have been programmed just to compete. We are highly trained to complete technical tasks, whether it is a skilled trades worker, architect or engineer, VDC specialist, project manager, or basically any other position you can think of, almost all education focuses on technical skills such as cost management, risk management, scheduling, and estimating. The most basic human functions of effective communication and conflict management are left up to each individual to learn on their own. The result of all of this is that the industry has become a transactional environment where in an attempt to achieve maximum effectiveness we focus only on what is required to get the job done and neglect how it gets done.

As expressed in the previous section, I believe this goes against the basic human needs of connection. The good news is that this presents an easy opportunity for you to accentuate this skill and differentiate yourself from others.

There are two key elements to expressing empathy toward others:

1) Reflection: along with our own experience and bias, we carry around mental maps for logic, which may seem counterintuitive to others. By taking a moment to understand where the other person is coming from or asking yourself what you would do in their position, you may be able to shed some light on what results they are trying to achieve and why.

2) Dig deeper: instead of asking the other party what they are trying to accomplish, try to dig a little deeper and ask why or try to determine this on your own. There is a theory made popular by Taiichi Ohno of the Japanese auto manufacturer Toyota regarding cause and effect; he suggests that the real cause can be determined by repeating the question "Why?" five times. I am not suggesting that you ask the question why in the same conversation 5 times in a span of 60 seconds – anyone with young children knows that this is not a very effective way to get answers and will just annoy the other person. Still, you can find ways maybe over a series of conversations or exchanges to drill down and find the real motives and incentives of the other person.

In doing the research described above, you are aiming for a middle ground and to determine what the other side is willing to concede on. The more you investigate it, the more you may find that there are common or alternative solutions to the problem and other opportunities to connect.

Humor – The Icing on Top

Humor is an underrated tool, especially in the construction industry. There is no denying that a well-timed joke can break the ice in a tense situation or create a positive and lasting impact on the listeners, strengthening your trust and likeability. However, it is probably one of the toughest skills to learn because it is the least inherent in people and the most difficult to teach. One suggestion is to find someone you work with who has a particularly good sense of humor and just take note of what they say, when they say it, and in what situation. A lot of the same jokes can be recycled, or similar jokes can be used in similar situations with others. One example of how I like to use humor is through storytelling before meetings, particularly ones during which we will be discussing a contentious issue. As you probably know, construction is always filled with problems to overcome, issues to resolve, and fires to fight. The bigger an item is and the longer it remains unresolved, the more people get involved and at a higher level so things can get serious. I try to arrive at meetings a few minutes early as it provides an opportunity to chat and connect with others before getting down to business.

Adding humor, particularly through storytelling, is a powerful connection tool, especially if it involves you personally and somehow connects to the meeting agenda or has some commonality with others in the group. In addition to improving your connection with others, humor can lighten the mood of the room allowing for improved communication and collaboration and reduced defensiveness. It will make the meeting smoother, more productive, and more enjoyable for all ... to the extent that meetings can be enjoyed!

The key here is that you do not need to become a stand-up comedian to be an effective communicator. Just understand how wit, sarcasm, and other types of humor can positively impact your relationships and your ability to connect with others.

Certainly, a well-planned story, joke, or comment can improve your relationships and others' perceptions of you, but it is also important to understand that the opposite can be true. There is a time to express humor and a time not to. An example of when to keep your comments to yourself is if the intent is to mock, make fun of or belittle a coworker or someone in your job site or office. Bullying is a real problem in the old construction world and most workplaces with modern standards have a zero-tolerance policy for bullying or harassment, though this does not stop some people from acting inappropriately. I believe that it is everyone's duty to stand up for those who cannot stand up for themselves. As an example of this, it is often considered a rite of passage for apprentices to be made fun of or mistreated. This is unacceptable and should not be tolerated. We all need to stand up for each other to make the industry a more welcoming place for all.

Another example of when it is not appropriate to make a joke is in light of a recent major event, which could include an injury, cause for a delay, someone being fired or laid off – anything that has a major negative impact on one or more people or the project. This is a cheap form of humor that will show that you do not actually care; it will negatively impact others' trust in you. You may get a couple of laughs here and there but overall, it is going to hurt you and is just not the right thing to do.

You will also need to learn when enough is enough. You do not want to become known as the person who just tells jokes all the time because construction can be a serious industry and rightly so. We are in the business of dealing with lots of money, lots of risk, and lots of people and they are all very busy. If you waste people's time, you will lose their respect. Another good tool to use here is to put yourself in other people's shoes before you make a joke. Think about how you would feel sitting across the table and having somebody else make that joke at that moment in time. I know we talk a lot about self-reflection in this book, but I do not believe that it is possible to spend too much time on it and most people, myself included, do not do it often enough.

4

Your Career Arsenal

Education Versus Experience

Many people I speak with ask me how relevant education is in construction. The answer depends on which facet of the industry you are in. In some professions like engineering or architecture, certain academic and industry requirements are required for you to practice as you need to apply the theory and knowledge that was acquired in school. In this case, education is extremely important. In the example of contractors, having an undergraduate degree may be a prerequisite for a position but when it comes to conducting the day-to-day work, you may use very little or none of the theories that you learned in school. In this case, the prerequisite usually serves as demonstrating a certain level of intelligence and problem-solving skills.

In many cases, and maybe more commonly in larger companies, you may find that more senior authorities have designations such as professional engineering titles, PMP designations, or GSC certifications. Certifications and licenses never hurt but are not necessarily required to conduct your job, especially as a contractor. John C. Maxwell said it best when he stated that titles and designations are like the tail of a pig: they sure look good, but it does not make the bacon any better.

Academic experience provides a good foundation and basis to springboard your career, especially in today's industry where most people have some sort of postsecondary degree and it is likely a requirement to enter certain parts of the workforce. This presents an additional challenge for

The Human Side of Construction: How to Ensure a Successful, Sustainable, and Profitable Career as an AEC Professional, Second Edition. Angelo Suntres.

those who are educated internationally where it may take some time and effort to get accreditation versus local requirements. Regardless of where you did your schooling, at the start of your career, you will have limited experience through co-op or internship opportunities. As a result, you will likely rely solely on your educational background to get started. While education is important, it is critical to remember that this is just to start of the journey, and you still have much to learn. I recall during my studies in mechanical engineering some of our professors telling us about how when we graduate, we would be trained professionals receiving six-figure salaries. Unfortunately, in my experience, this could not have been further from the truth! Yes, post-secondary degrees like engineering, architecture, and other heavily technical fields require high intelligence and hard work to attain but it is important to acknowledge that the majority of this information is learned from a book and not lived experience. True value lies in acknowledging that while you may be strong in technical skills, there is a lot to be learned about how to apply them in real-life situations. This is best summarized in the following story from a fourth-year business professor whose course I took during my studies. On the first day of class, the professor said that this is the most important course we will ever take! He went on to say that on our first day at work, nobody is going to ask us to calculate the heat transfer through an infinite sphere or ask that we solve a triple integral. We all laughed, but we had no idea at the time how true it would turn out to be.

Completing a degree like engineering or architecture can and should be worn as a badge of honor; it shows you have what it takes to put in the time, effort, and work required over a long period of time to achieve something great. This is actually, in essence, a great analogy to a construction project and its significance should not be overlooked regardless of your level of experience.

Even if education is all you have at the start of your career you can still apply it to hands-on tasks for the job you are applying to. You may need to get creative but try to tie your specific knowledge to specific tasks in the job description. This will show that you have relevant experience even if it is just education. The fact of the matter is that when you are fresh out of college or university, there are likely hundreds or maybe thousands of other people who have graduated at the same time with the same credentials. So, you need to find a way to differentiate yourself.

To summarize, education is very important, but it really is just the starting point where the door opens and provides opportunities for further learning. So, remain open and willing to take on new challenges to apply theory and develop new skills through experience.

We find ourselves at a very interesting time in the construction industry. On one end of the spectrum, we have people from the "baby boomer" generation who are at or close to retirement age and have a wealth of knowledge and experience. Typically, these workers rose through the ranks from field-oriented positions, and most had little to no formal education beyond secondary school. This is not to say there is any shame or anything wrong with this choice; it was just a different time 30 years ago when they were entering the trades or workforce. On the other end of the spectrum, we have new graduates who are highly educated but inexperienced. They are full of life, potential and technical knowledge, but they have not quite figured out how to put it into use. Then there are a few people in the middle of the spectrum usually found in middle management-type roles. The result is a huge generational gap in the workforce, which poses some issues in the way the team communicates and operates.

There is no substitute for hard-earned experience, and you can learn from both the good and the bad. Try to get as much exposure to different aspects of the business as possible including estimating, project management, site supervision, business development, and finance. You do not need to work specifically in any one of these departments for a certain period of time but ask to get involved in different capacities such as assisting with a certain task. It affords a great opportunity to learn and meet people to expand your network and skillset.

Ultimately, it is your experience that will determine your overall success in the construction industry. The good news is you have a certain level of control over what experiences you have as they are dictated by the habits, decisions, and connections you make with people along your journey. In the past, many people started on a certain career path and did not think twice about the direction they were headed as it was typical to have worked for the same company for 20 or even 30+ years, gradually climb the corporate ladder, and get a gold watch and retire. Simply put, these days it is just different. For many reasons, the merits of which are debated, it is more common now to "job hop" from company to company in stints as short as six months. These choices will quickly bring you different vantage points, a more rounded experience, and incremental increases in salary/benefits but you also need to consider your long-term goals. To advance into upper management, for example, in many companies, it takes time to build trusted rapport with the right people, which is negatively impacted by frequent job/company changes. Of course, this all depends on what your goals are for your career; not everyone wants to be in senior leadership and that is perfectly fine. No matter what your path and goals are, the principles of this book will still help you get the most out of your decisions and maximize your career.

How to Leverage Everything You Have

The previous sections covered experience and education and their individual impact on your career. In this section, we will discuss how to combine both to maximize your assets and boost your career. If you were to list out all your academic and work-related accomplishments onto a resume or CV, it would create a snapshot from the past – a wonderful opportunity to demonstrate proven things that you have accomplished. Of course, in this case, both education and experience are important to include but some people struggle with what information to include and what to leave out. There is a delicate balance between too much or overly specific information and not enough or overly general. It is not just about having the right education or experience but how you demonstrate it on your CV and through the interviewing and hiring process, there is only a small period of time to impress people. The best way to highlight educational experience is not simply listing the degree and the school that it was obtained from, but rather getting into some specifics such as specializations, certain courses, or any type of honors received as a student that would distinguish you from others. You may also have had other experiences outside of academics during the course of your studies such as volunteering or leading organizations. This is an excellent way to demonstrate that in addition to academics, you took on additional responsibilities and developed or applied business or people skills concurrent to your studies.

If you have some level of experience in the workforce, it is good to tie it to your education and show how all the pieces fit together in the grand scheme of things. As an example, if you studied mechanical engineering in school and started working on design-build projects, then you were able to apply your engineering technical theory in practice while complimenting this with the building side. More emphasis should be put on the experience you have rather than education because it is easier to prove and demonstrate your abilities from real life situations. Obviously, there will be more examples if you have been in the workforce longer but chances are even if you are fresh out of post-secondary you will have some examples of how these skills have been applied on site or in the office. Many employers especially in the construction industry are more concerned with how you can apply the skills and knowledge you have acquired than with the degree or certifications that you have obtained.

Focus on your best qualities, how they are demonstrated in your experience and supported by your technical studies. By doing this, you are providing real-life-supported examples so you are really capturing the big picture and setting yourself up for success.

Regardless of what stage you are at in your career, you can always strive to do more or do better if you so choose. Many people in construction want to keep active in the industry and their community even into retirement years. Some of the best leaders you will find are on a continual path for constant improvement and crave action. While the previous section focused on what you have accomplished to date, it is important to note at this point that what has happened to you in the past does not define your future. If you have come to the realization that you may be ready to pivot in your career that is perfectly fine, and it does not mean that all the experiences and connections that you have acquired have gone to waste. In fact, if you give it some thought, you will find that there are many ways to transfer the knowledge and experience that you have earned even if you are considering changing careers completely. If you are just starting out and have just completed a post-secondary degree, you are full of potential but light on experience. If you are toward the end of your career in a more senior role, you are full of experience and impact but likely low on potential. It is important to understand where you are at on your journey to set realistic goals and expectations to avoid disappointment or frustration. I have spoken with many younger employees, who feel a sense of frustration when they feel that they could take on more responsibility and have a bigger impact within the team, company, and industry but are being held back by perceived lack of experience or poor leadership.

The most difficult part about being young and ambitious is that young is associated with a lack of experience and while ambition and potential are particularly important early on in your career, experience is king in construction. Unfortunately, it takes time to acquire the necessary experience to get more responsibility, get promoted, gain respect and trust, etc., and this does not happen overnight. To provide yourself a successful and sustainable next step, you will have to build a stable foundation. Another way of explaining this is that in order to manage people or gain more responsibility, you must have a firm understanding of the nuts and bolts and day-to-day activities that are required to do that job. For example, before you can become a Project Manager, you need to understand the basics of a Project Coordination or else you will not be managing projects for long! I am not encouraging anyone to become a micro manager – in fact, that is one of the worst things you can do for your career – but you must know the basics to be able to spot red flags or when somebody is trying to pull the wool over your eyes.

If you are submitting your CV or interviewing for a role, expressing what you think you are capable of will not in itself be convincing unless you are an exceptional salesperson. Ultimately, you will need to back these words up with proven action. For example, if you are a Junior Estimator seeking advancement and you recently completed a smaller estimate on your own including contacting vendors, completing the takeoff, assembling the bid sheet, etc., then in discussing future opportunities with your manager, you are able to prove what you did in detail with a real-life situation and express interest that you would like to work on larger or different types of projects. When you can demonstrate that you perform your current tasks efficiently, it proves that you are ready for more. You must find a way to tie your experience to your potential to really drive the point home to improve the chances that people will support you in your advancement.

Learn From Other's Mistakes

In the construction industry, as in life, learning from others' mistakes is a valuable strategy for personal and professional growth. It is a shortcut to gaining wisdom without having to experience the pain of failure firsthand. Consider the construction site as a classroom where every project, every task, and every interaction are a learning opportunity. Observing others, especially those with more experience, can provide invaluable lessons that can save you time, frustration, and embarrassment. This can work in two ways: first, consider the example of a seasoned project manager handling a crisis, which can show you effective problem-solving under pressure. Conversely, witnessing a poorly managed situation can teach you what not to do. Both are very important and usually the latter provides the most powerful and memorable lesson.

Learning from other's mistakes is about understanding the situation, the decisions made, and the outcomes, then reflecting on how you might handle it differently. It is about fostering a culture of continuous learning and improvement, where mistakes are seen as opportunities for growth rather than failures. Remember, nobody is perfect and even the most experienced professionals make mistakes. The key is to have an open mind, be observant, and be willing to learn. When someone else makes a mistake, do not just dismiss it as their failure. Ask yourself, "What can I learn from this? How can I apply this lesson to my work?" It is ok to make a mistake – as long as you learn from it – but never make the same mistake twice!

Moreover, do not limit yourself to learning only from people within your organization. The construction industry is vast and diverse, with countless professionals who have faced similar challenges. Engage with people working in other companies, trades and industries through forums, attend seminars, read case studies, and connect with peers. These can all provide insights into common mistakes and how to avoid them. Oftentimes, the best perspective is brought from a fresh set of eyes which are not accustomed to the status quo and are able to think more critically about situations and results.

Finally, remember that while learning from others' mistakes can accelerate your learning curve, making your own mistakes is inevitable and not necessarily a bad thing. Mistakes can be painful, but they also provide potent lessons that stick with you. As the saying goes, "Experience is the teacher of all things." So, do not fear mistakes, but strive to learn from them quickly and turn them into a positive for the advancement of your career.

In conclusion, integrating others' experiences into your own can help you navigate the complex world of construction more effectively. It allows you to build upon the collective wisdom of those who came before you, accelerating your career progression, and helping you avoid common pitfalls. So, keep your eyes open, be humble, and turn every mistake, whether it is yours or someone else's, into a stepping stone toward success.

5

We Are All Salespeople

Find Your Value Proposition

Whether or not you like it, you are a salesperson. If you work in the business of dealing with other people, there are constant transactions going on back and forth where something is bought and sold. There may or may not be money exchanged but it always involves social currency. Social currency can be defined as "influence on social networks, online and offline communities, and the degree by which your business is shared by others" and is based on your relationships. Much like financial investments, if you have many diverse, rich relationships with different people, you have a strong social currency. You may even have only a few relationships with very high effectiveness and still possess strong social currency. The good news is you can increase your social currency with minimal effort by being nice to people and aiming to bring them value. Connect! Even in the construction industry, where you may have a heavily technical role, you are still selling two things:

1) Yourself (your personality and ability to connect)
2) Your service (what you do for your profession)

Notice the order in which these two are presented. Many think that connecting with someone in a professional setting is purely based on your service; however, going back to previous examples, we know that you first must connect on a human-to-human basis before adding on the complexities of money and construction or other businesses. Therefore,

The Human Side of Construction: How to Ensure a Successful, Sustainable, and Profitable Career as an AEC Professional, Second Edition. Angelo Suntres.

it is important to keep in mind that, while the commonalities that bring you together initially may be your occupation, to establish a deeper, more meaningful, and more beneficial relationship, you must connect on a personal level. This is done by selling yourself – your personality and your value-add. By reflecting on your experience, strengths, and weaknesses, or conducting the self-exploration exercise in the appendices, you should have a strong idea of what personal attributes you excel at and can leverage to build the bonds required and establish connections to ensure your success. Remember that when selling yourself, there must be true value added and reciprocity in the relationship, so it is important to focus on genuine connection or else you come across as a used car salesperson. Find strengths in the other person that you can connect over (commonalities) or positive traits that the other person possesses that could supplement some of your areas in need of improvement. This expression of humility will make the other person feel more powerful and help create trust.

I do not suggest that you point out where others are personally deficient and where you can assist but rather focus on operational or business-related needs to which you can bring value. An example of this could be the following conversation between a Project Coordinator and their Project Manager:

> PROJECT COORDINATOR: *I heard you mention in the meeting yesterday that you are swamped with millwork shop drawing reviews. I know you are super busy with so many things. I coordinated the millwork package on my previous project and would be happy to help if you need some assistance.*

Bonus points for complimenting and offering help at the same time. Try not to come across too strong or "suck up" to the boss as this could work against you. As with everything, there are many fine lines! Don't be afraid to try different strategies in different situations. Little mistakes make the best lessons.

Let's face it, your service in construction and most other industries is a commodity. Unless you work in a very small niche in a very specific sector, you – at a fundamental level – are replaceable based solely on technical skills and job function. I am sorry to be so blunt but it is true. That is why you need to focus on combining your service with your personality and relationships to develop your personal value package. The real value added in organizations is people, not positions. Remember that people do business with those that they know, like, and trust. So, if you take two workers with equal ability, but where one is well respected, liked, and trusted, the other is not a team player, not personable and not reliable, who do you think will get promoted? Who do you think will keep their job during layoffs? The answer is obvious when presented in such a straightforward fashion but now is a great time to ask yourself, which worker are you? If you're the latter, remember that it is never too late to change ... but start now! Right now!

Surely, if you are a super-connector but can't do the basic functions of your job then you won't last long (even if some people do last longer than they should sometimes!). Conversely, if you are excellent at your role but not very social and isolate yourself from others, you will experience a certain level of success but may not make it to management or beyond. Again, this comes down to personal preference. Not everyone wants to make their way into management or senior leadership but the fact that you are reading this book makes me think that you do, so let's keep going!

It is not a competition about who can do their job better or who has more friends; you need to find a balance of the two. Comparing yourself to others in either category will only set you up for failure. Measuring yourself by others' standards will never result in fulfillment or satisfaction since there will always be someone who appears to be doing better than you. This is another reason why it is important to conduct the self-assessment to reverify what you will be measuring yourself against, so you avoid falling into the trap of trying to keep up with someone else.

Now that we have discussed and acknowledged the importance of selling ourselves and our value-add both personally and professionally, let's look at some ways to express it and connect with others.

How to Sell Yourself

There is a lot of talk these days about building your personal brand. Your personal brand is basically what people picture when they think of you. When you speak to most people, especially when first meeting them, inevitably they will ask what you do for a living. Most people answer with their job title and where they work: "I am an Estimator with a general contractor" or "I am a Project Manager with a civil company." Boring! How many estimators, project managers, and site superintendents do you know? There are tons .... Focus on what makes you different ... what makes you YOU! Remember that people connect with people, not roles or titles in a company. Sure, this standard answer may lead to more questions and start a conversation like, "What types of projects are you working on?" or, "How long have you been in estimating?" Still, this does little to differentiate you from the next worker with the same title and experience. Your personal brand could be the difference between being remembered and invited back or being left in the dust.

Make your introduction memorable. You may have only minutes to connect with a person before they are off to the next meeting or phone call. An example of a memorable answer to the question of what you do is, "I am a construction leader with a passion for leadership, changing the industry one person at a time." Yes, this is my tagline ... but how can you hear this without wanting to learn more? Obviously, you must have some sort of support for your statement; you might look foolish if you are caught without an explanation for a beautiful, intriguing opening line.

The good news is that it is easier than ever these days to build your brand. The best way to do it is to get out there and make positive connections with people. The biggest impact will come from face-to-face interaction but this limits your audience in most cases since some key people required on your journey will not be easily accessible. Social media is a great way to supplement your personal brand; LinkedIn is a perfect example of a professional online platform on which you can easily create a profile, connect with anyone in the world (provided they have an account and accept your connection request), view and get ideas from thought leaders in your industry, share your own thoughts, and make new connections. Sounds great, doesn't it! This book is not intended as a resource to build your LinkedIn account, but it is actually very easy to create and optimize your profile for connecting with people. There are lots of great free resources online for this, such as YouTube.

Now that you are all set up to sell your personal brand, don't forget about your actual job! In the next section, we will cover what to focus on for your day-to-day tasks to complement your personal brand without sacrificing the work you get done.

Be awesome at your job! This might sound like simplistic and useless advice. Everyone wants to be good at their job, right? Well, not if you are miserable and underappreciated. Have you ever heard of quiet quitting? This is when people stop going above and beyond in response to being underpaid or undervalued in their position. I totally agree with, and support, being paid for your worth. If you are underpaid or mistreated by your employer, absolutely it is time for a change. Life is too short to waste your time with someone who does not value and respect you. I know it is not that easy but by this point, you should have a good idea of what you like doing, what you are good at, and what you can make money doing. If this equation is not adding up in your current role, start looking for something better if you haven't already. I am only covering this topic in this section because if you are not happy in your current role due to external factors, then any changes you make to your personal skills will have little impact and may further discourage you. It is very difficult to be awesome at your job if you are not happy doing it.

Generally, being awesome at your job has two factors:

1) How effective you are in completing day-to-day tasks
2) How you impact others while doing (Spoiler alert: #2 has a bigger effect than you may think)

Consider an example comparing two types of workers with equal technical abilities. Worker #1 completes all tasks on time but makes no effort to build rapport with others. He keeps to himself even in collaborative settings. Worker #2 is moderately reliable, but everyone enjoys working with them. She is such a team player that people seek advice from her for work and personally related items because they know and trust her. Based on the information provided, which one is more likely to succeed? When you think objectively about this example, the answer seems obvious. On paper, the person who is the most reliable and effective should come out on top but they do not; why? People are not machines that merely produce an output. In interacting with others, people gravitate to those they like, respect, and trust. Now, I am not advocating for people to be promoted solely based on their friendships or likeability. If you cannot complete the basic job function in your current role, there is no way that you will get any meaningful or sustainable advancement. However, and this is an extremely important point, it is not what you do but how you do it. The actions you take are more likely to be remembered by the impact you made on others. As Maya Angelou said, "People will forget what you said, they will forget what you did, but they will never forget how you made them feel."

6

Where Has the Trust Gone?

Why There is Low Trust in Construction

If you speak to anyone who has been in the construction industry for five years or more, chances are they have been burned by someone or something at least once in their career. There are a lot of people out there who are only looking out for themselves and their profit and will sacrifice anything to get it. The construction industry has many good, honest people but a few bad apples have marred its reputation with the general public and many owners/clients. This is unfortunate as it impedes the development of the most important element of any relationship – trust. It is almost as if contractors are starting with a negative balance in the trust department to make up for mistakes or greed from other people in the industry. Therefore, it is important to acknowledge this issue upfront when dealing with a new customer, subcontractor, vendor, or coworker. Many people come into new projects, negotiations, and relationships ready to fight and expect to get hurt.

Take the example of an estimator taking prices from an equipment vendor on a project. Vendors and subcontractors typically have pricing tiers when closing jobs. If you are a loyal customer with high trust, you are tier 1 and you usually get the best price. If they do not have much experience with you or have low trust, your price will be higher. Some companies go as far as asking for the best price at closing, only to negotiate an even lower selling price once the project is awarded. This may seem like you are getting a deal, but at what cost? When there is an issue

The Human Side of Construction: How to Ensure a Successful, Sustainable, and Profitable Career as an AEC Professional, Second Edition. Angelo Suntres.

on site that requires extra effort, how likely is the vendor/sub to go out of their way to help you? How will this impact future bids? I have heard that some vendors/subs build "fluff" into their quotes depending on who their customers are because they know they will be getting a "haircut." I prefer to focus on making money for the rest of my career, not one job, so I choose the relationship over the quick win. If more people thought this way, at all levels, I believe the industry would be able to restore trust and building would become easier and more profitable ... but that is my utopian dream.

I am sure that you can reflect on some negative experiences that you have had; you have probably been on the giving end of your share as well. I know I have. I refuse to believe that it is "just part of the job." Again, it is never too late to change. If we all change the way we think we will create a new norm that is less combative and more collaborative.

As previously discussed, prior to conducting higher levels of function in any relationship it is important to form a solid foundation based on trust, honesty, integrity and mutual respect. Much like a building, if a lot of force is applied to a foundation that is not structurally sound – either by faulty design or deficiency – it will eventually get cracks or even experience catastrophic failure. It is no different when it comes to relationships. If you neglect the basics of human connection and start to add the complexities of financial requirements, contracts, and the technical requirements of construction, you are bound to have issues.

This is another contributing factor to the lack of trust in the construction industry. A lot of focus is placed on training in technical skills such as engineering, design, estimating, project management, etc., with little emphasis placed on personal skills such as effective communication, conflict resolution, and the ability to have difficult conversations or make difficult decisions. If you do not have a strong relationship with someone – particularly in the trust department – and you start to apply the pressure and forces associated with solving technical problems, negotiating large sums of money, and conflict, there are going to be struggles and challenges that will be difficult or impossible to overcome. Construction by nature is filled with problems; this is part of what makes it such a great industry. If you are up for a challenge, there is always something to do and if you gain a reputation as a problem solver, you can go far in your career. Understanding that there are always going to be fires (conflicts, disagreements, and interferences) that need to be extinguished daily, it is wise to equip yourself with the tools required to mitigate risk, resolve conflicts, and find solutions. The traditional training in construction does wonders for specific problems that may arise during a project like coordinating trade interferences on site, submitting or responding to RFIs, or conducting a shop drawing review/transmittal. These are all separate but specific tasks that are critical to completing your job, but the basic principles of human connection are required to do all of these effectively in any situation!

In summary, acquiring technical knowledge to solve specific problems is important, but will be most effective when applied with the general knowledge of effective communication and interpersonal skills. This is what will enable you to demonstrate your effectiveness, improve your relationships, and advance your career.

How to Get It Back

In a perfect world, when you initially meet someone, a certain level of trust would be inherent and either increase or decrease based on your interactions with that person. The unfortunate reality is that in most cases, you will be starting with zero trust and will need to prove yourself to earn it from others. As described above, many people have been hurt by others in the past and have their defences up to avoid the same thing from happening again. The best way to combat this is to operate consistently with the key relationship tenets in mind: honesty, integrity, and trust. Even if you encounter someone who has low trust and is argumentative, responding with the same will only make things worse. It may take some time but this is where consistency is key. Anything else will only perpetuate the cycle of problems we are facing in the industry. This is easier said than done, especially in the beginning when you may feel that you are not making any headway or gaining any ground in the trust department. There may be a point where you realize that it is not worth continuing the pursuit, but you should only consider this after a constant sustained effort.

The length of time that it will take to gain trust depends on the person you are dealing with, and the amount of trust involved. If you can, determine what may have happened to the other person in the past either by asking them directly or through others to see if this is something you can address directly. They may have had a bad experience with someone else from your company. Unfortunately, this is often the case, especially among large organizations where people tend to paint the company with a broad brush because of their experience with one person. The larger an organization grows, the more likely it is that you will have a bad apple in the mix. This issue is especially problematic in an industry as decentralized as construction where each project is essentially a mini company that operates peripheral to a main office. If you find yourself in this situation, it can be addressed with a quick reminder of how the previous person's actions are not a representation of the overall company, that your philosophies are different, and that you are there to help them with their problem.

Another example is if the other person had a bad experience with somebody in a similar role to yours. This is common between owners and contractors where owners may have been taken advantage of on the previous job regarding change order pricing or other types of negotiation where they were treated unfairly. Again, this is worth addressing directly by having a conversation and stating that this is against your company's or your personal philosophy and emphasizing that you are there to help them. Keep in mind your place in the organizational structure; conversations like these have more meaning when they come from senior levels as they are typically more aligned with the company values and have authority over major decisions. You may need to get support from your manager to quell any worries or hesitations so you can start fresh in the trust department.

In the previous section, we covered ways to build trust in the construction industry. It is equally important to understand what not to do to avoid common pitfalls and perpetuate the outdated construction stereotypes. The role of everybody in the construction industry, whether you are an owner, contractor, site supervisor, quality control, etc., is to complete the project safely with minimal cost. On the contractor side, minimal costs will maximize profit. On the owner's side, minimal cost will reduce expenditures. The same is true for subcontractors and vendors, so why is it that construction has become so combative? In all aspects of the industry, there is a silo effect that creates an "us versus them" mentality. The result is poor communication leading to poor coordination, further leading to poor execution. Poor execution has a direct impact on productivity and profit. Is it reasonable then to assume that if we work on breaking down these walls, coming out of our silos, caring more about, and communicating better with each other, then projects will be completed quicker, easier, and more profitably? Or, is that too logical?

The most important thing that you can avoid doing that is sure to negatively impact trust is to not trust others. I choose to believe that everybody has good inside of them, they just need to be given the opportunity to prove it. It is up to each one of us to approach situations with an open mind and no hidden agendas and expect the same in return. Of course, this will not work with everyone you meet and you will have to know when enough is enough, but it is always advisable to give the other person a chance.

The challenge here is the misconception that the construction industry is difficult and so are all the people in it. In my experience, I have met some tough cookies but mostly find that people are generally kind and willing to help out if you show them a bit of respect and appreciation for the expertise that they bring to the table and provide the opportunity for them to share.

7

Dealing with Toxicity

Workplace Warning Signs

Over the recent years, and mostly due to the COVID-19 pandemic, the necessity to transition quickly to remote work and subsequent return to the office created a lot of focus and conversation around work cultures. The term "toxic" usually refers to the type of workplace that is not balanced and focuses on results and profitability over valuing and appreciating employees. This usually looks like unrealistic expectations of work including long hours for little pay. What some view as nice perks to have in the office like ping pong tables, pool tables, free food, and rest areas others find as underhanded ways for employers to keep people in the office for as long as possible to get the most out of them. While there is no one-size-fits-all approach, it is important for you to get a strong understanding of what your motivations are and what you expect out of your company in return for your time and effort.

Here are some signs that you may be in a toxic work environment.

1) You are overworked: A classic sign of a toxic work environment is one where you are required to put in unreasonable hours to accomplish your tasks. In this case, this should not be confused with people who have a hard time saying no and take on more than they can achieve. It is important to do your job well and there will be times, especially in the construction industry, when you will need to work longer days, evenings, or weekends. Still, this should not be the norm. Work is

The Human Side of Construction: How to Ensure a Successful, Sustainable, and Profitable Career as an AEC Professional, Second Edition. Angelo Suntres.

only a part of everybody's holistic life and there are other aspects that you need time and energy to experience and appreciate.

2) You are underpaid: Another common sign of toxic workplaces is low pay, especially coupled with the dangling carrot of incentives like bonuses or salary increases. Everyone deserves to be paid their worth, though this can be a subjective topic. It is useful to know what other people in similar roles to you are making to help inform your decision whether to stay and tough it out or look for something new.
3) You don't feel valued or appreciated: If your management or leadership team treats you strictly as a resource to get a job done, then you are likely in a toxic workplace. Examples of this could be feelings of shame or guilt for wanting to take time off, even though you are entitled to certain vacation time or bringing up a personal concern such as dealing with mental health issues and getting little or even negative feedback.

These are just three common examples of behavior that exist in toxic workplaces; you can probably think of others. The overall feeling in these types of workplaces is that you give more than you receive. This is not sustainable for a long-lasting, enjoyable career and not the positive type of environment in which you can best enhance and apply your leadership skills.

It is not uncommon to feel trapped in these types of work situations. If you feel that you are in a toxic workplace culture, the first thing you can do is assess how you got there. Depending on your background and experience, you may have taken the job out of necessity. You may even have had difficulty finding this role and needed to take it to provide the basics for yourself and or your family. If that is the case, it is not as easy as finding something better as many people, especially on social media, will suggest. If you find yourself in a position where you are giving more than you are getting in return but for whatever reason, you can't find something better, there are some ways that you can protect yourself to make your job more enjoyable or simply bide your time while you find something different.

One skill that you can learn is how to say "no" tactfully. The first step here is knowing your limits and what can reasonably be done in your workday. The key here is to not simply say "no" and leave it at that but rather to provide some explanation as to why you cannot take on additional work. For example, if your manager comes and assigns you a new task, you can reply with, "I could take this on, but it will have an impact on other activities I am working on. Can you please help me prioritize what needs to get done?" This is a simple way of explaining that your plate is full and by adding other tasks, something else will be affected. This also puts the onus on your manager to determine which items are a priority, which will help in setting expectations and avoiding disappointment and frustration. Notice how this process involves effective communication!

An important thing to keep in mind if you find yourself in this situation is that nothing is permanent. Even if you feel stuck and you need to remain in the position for a time until something better comes along, stay positive and try not to let how others make you feel impact your self-esteem and self-worth. This is especially true if you are not seeing the results you would like or expect out of your career; it may not be because you are doing the wrong things but because you are in the wrong environment.

Personal Traits to Avoid

Just like positive relationships and experiences will lift you and help you succeed, negative relationships can do the exact opposite. In the last section, we covered signs of a toxic work environment and ways to manage or make a change. In this section, we're going to discuss the same topic but at a more micro level when it comes to different personalities that you will encounter in your career.

In organizations of any size, you are likely to find toxic people despite the efforts of human resources, people, and culture or senior management. Your organization may have the highest standards and values for treating people with respect and establishing good healthy relationships but somehow, people who do not align with these values and beliefs permeate their way into the workplace. I am sure you can think of someone from your past that fits this description. Their presence sucks the life out of you. This type of behavior is contagious and can spread like cancer through an organization. After spending time with them, many will feel exhausted both mentally and physically. Having said that, I believe that most people can change if they want to. Many who may exhibit toxic traits probably have no idea that they are doing it. In this case, they may welcome a conversation identifying their behaviors as a way of trying to improve their relationships and your workplace. If you have a rapport with the person, you may be able to initiate this type of conversation but, at the end of the day, it is the responsibility of management to ensure people like this are addressed. Things rarely get better on their own and if it is not addressed, it can have lasting effects on relationships, careers, and organizations.

It is important to be able to identify this type of person and how they can affect your career. In addition to mentally and physically bringing you down, they could affect your other relationships if others start to associate you with them. I know this may seem counterintuitive to other points in this book; after all, all relationships are important, right? While I believe it is important to be kind, try and bring value to all you come across, the reality is that not everyone wants that and to give it in return. You must use your better judgment to make the call and remember that all relationships play a role in your support system and ultimate success. Choose wisely!

Depending on your situation and role, avoiding these types of people may be difficult or even impossible. So, it is important to understand how to cope with them. Like with toxic environments, you may not be able to avoid these types of people, especially if they are your coworkers or managers. If you are required to interact with someone like this on a day-to-day basis, it can cause stress or anxiety before, during, and after the interactions. You cannot control the people around you and how they make you feel but you can choose how to react to it. If you are stuck with the person, learning to manage your stress or anxiety can help maintain a healthy relationship with yourself and decrease the impact of the person on you. There are many ways to deal with stress and anxiety and the level of effectiveness varies from person to person and the situation you are in. Physical activity is a big one for lots of people; simply getting up and moving around increases blood flow and will help alleviate some negative feelings. Going outside for fresh air is another good example of physically removing yourself from a situation and getting some increased oxygen. Sometimes though, you will be in a meeting where it is not appropriate to get up and move around or excuse yourself. In this case, an inconspicuous tool that many find useful is breathing exercises.

Dealing with stress outside of work is very important. I am speaking from experience when I say that it is easy to bring home work-related stress and project it onto your family, spouse, kids, or even your dog, without realizing it. Stress is part of life and unavoidable at times. So, an important part of your success and enjoyment is finding a release for this. Examples could include going to the gym, playing an instrument, or talking to a therapist.

The strategies mentioned here are just a small number of techniques and outlets. Do some of your own research, try different things, and see what works for you.

Another example of effectively handling toxic relationships is learning how to limit exposure to only the amounts required. Often a toxic person will talk incessantly about a certain point even if the conversation started on a relevant topic. Try to keep conversations to a strict agenda to avoid any tangents. Mention that you don't mean to be rude but you have a lot of work to do and have to get back to your work area.

Time and energy are your critical resources for a successful career. The key here is not letting other people's negative resources steal yours. Avoid these relationships where possible. And where you can't, learn to minimize contact and limit exposure.

8

Critical Traits to Embrace

Care for People and They Will Care for You

I hope that the information we have covered so far has helped you discover, or at least consider, yourself including how you operate, what brings you joy, and career goals to strive for in the future. In the next sections of the book, we will cover some skills to consider adding to your arsenal as they have proven to help many people in their careers.

Kindness, along with empathy is absolutely the most underrated skill that you can exercise to connect with others. Remember that leadership is about connecting with others and adding them to your circle of influence. To me, there is no better way to establish a connection with someone than to simply be kind to them. Show care, show that you want to help, smile ... these are simple things that usually get passed by in many situations. Despite how simple these are, they will have a profound effect on how people remember you. Remember that most of what happens when interacting with people – especially in a first impression – is happening at the subconscious level. By showing kindness to someone, you impress upon them that you are a safe person and can be trusted which, as previously discussed, is a fundamental building block in any relationship. Almost all the professionally powerful people I have met in my life were surprisingly warm, cheerful, and helpful, even after meeting only briefly.

The good news is that this is an easy win in the construction industry, where rough, tough, and sometimes rude behavior is stereotypical. From this lens, kindness may be viewed as weakness but this is not the case.

The Human Side of Construction: How to Ensure a Successful, Sustainable, and Profitable Career as an AEC Professional, Second Edition. Angelo Suntres.

You can stand up for what you believe in without being rude, you can prove your point without getting upset, and you can win arguments without yelling. The best arguments are won by outsmarting your opponent, not through intimidation or abuse. All people, even children, respond the same to physical or emotional threats. Here is a perfect example:

A group of workers is socializing during working time when the boss comes in and starts yelling and swearing at them to get back to work. The workers are likely to get back to work at that point, but what happens when the boss leaves? Now take the same scenario but apply a kind and caring approach. When the boss comes in, instead of screaming, he encourages socializing during breaks and mentions that when work time is affected, so are productivity and profit, not to mention the Christmas bonus. By offering an intrinsic award, they improve the chances that the workers will make the right choice because it is the right thing to do (internal), not because they fear punishment or harm (external). The former will work all the time, the latter only works when you are present.

Be kind but take no crap. The key to being kind without expressing weakness is expecting the same treatment in return. If you are nice and people take advantage of you, it will not be effective and could work against you. Being kind should not be confused with being a people pleaser, who avoids conflict and makes everyone happy regardless of the cost. It is important to know what is right and important to you and focus on the main objective – maximizing your relationships and success – though there will be situations where you need to concede. You can't always win!

So, how do you deal with someone who is difficult or mean? Be extra kind. Do not sink to their level or they will have already won despite the outcome of the conversation or situation. This can be extremely difficult, especially if you get a special person, who provokes or insults you and may even call you out for being weak. Exploiting weakness is a perfect example of a toxic workplace or personality. Trying to make someone feel inferior because of the way they're acting is an attack on your self-worth and respect and should not be tolerated. Remember that kindness is not weakness. It is a superpower to remain calm in difficult situations and is an important attribute to possess. Not only will this help you in many construction situations, but it is a great asset that leaders will notice and will help you get noticed.

Stupid people are everywhere. So, you are going to find yourself in undesired situations. In my experience, what helps is telling myself that the person is being irrational and trying to goad me into an unnecessary conflict. I remind myself that I do not need to stoop to their level to beat them at their own game. Take the high road. Again, it is possible to stand your ground and prove your point calmly with effective communication and a healthy dose of wit. Staying true to your plan will cause you less stress in the long run. And if you are not making any progress, it is ok to agree to disagree at the moment or plan to discuss in a different setting – many people act differently alone than in front of their peers as they need to uphold their reputation. Depending on the severity of the event, you can even escalate to get a resolution.

There are always options that can solve problems but the key takeaway here is to lead with kindness and care and expect the same in return. If it is not returned, then you need to evaluate whether or not pursuing that relationship is worth the stress or effort.

Change is Constant – Adapt and Overcome

In today's world of ever-expanding technology and innovation, if you are not able to adapt to change, you will not thrive, and you may not even survive. There has never been a more exciting time to be in construction. We are starting to experience a paradigm shift as the next generation of leaders take the reins in organizations and sites. There are many changes that are occurring but for the purposes of this section, we will focus on two: humans and technology.

There is a revolution occurring that is starting to acknowledge what I call the human side of construction. One key element of this is the emphasis on soft skills, like those taught in this book, that have traditionally been lacking in educational systems and onsite training. The next generation of construction leaders are more intuitive than their "boomer" counterparts and rely more on feelings rather than cold hard facts. Perhaps this is due to more of an education-based career than experience. These newer methods of communicating and decision-making rely heavily on human connection, which no one has been trained in! This presents a huge opportunity for anyone willing to invest the time and effort to master these skills, which includes you because you are reading this. So, congratulations!

Another example of the human element is mental health awareness. Even as recently as five years ago, very few people were speaking about or researching the topic of mental health, especially in construction. Now it is a common topic covered in safety and other organizational meetings at all levels of industry. There have been such tremendous efforts in the construction industry to mitigate and eliminate physical safety risks through the use of PPE and safe work practices. The results have been phenomenal! It has never been safer to work in the industry. The result is twofold: (i) Safe work is productive. Physical incidents cause work stoppages, investigations, and ripple effects which all equal delays. (ii) Safe workers are happier. When you are equipped with the tools and training to do your job safely, you don't have to worry about getting hurt. So, you are happier. This will translate into increased performance. All of this leads to more productive, profitable work.

The same principles are starting to emerge in mental health and safety and, when you think of it, the same results are likely to occur. Using the same examples as above: (i) Safe work is productive. When you are mentally healthy and safe, meaning that psychological issues are addressed, you are less likely to make errors that could lead to incidents. (ii) Safe workers are happy. When you feel safe, expressing your true self, whether black, white, male, female, gay, trans, etc., you are going to be happier. Again, this will translate to increased productivity and profit.

The second major point of change happening now is technology. I am sure that you have probably read reports and studies on how construction is one of the slowest industries when it comes to adopting technology and innovation. This does not mean that there is no technology or potential for advances in the industry but rather that there has been a hesitation to adopt what is available. Just like the next generation of leaders is more intuitive and communicative, they are also more likely to take risks on technology and innovation; after all, they do not have as many years of experience to get entrenched in their way of thinking ... and they were born with the internet! When it comes to adoption of technology, the question is not when, but how and what. There are many services and products out there now that can improve your organization, but they have to align with your current processes, values, and – most importantly – the people, who will be implementing and operating the shift.

Sounds easy, doesn't it? The challenge lies in being adaptable and embracing change. Many people struggle with this, and it is more prevalent in the construction industry because the schemas, attitudes, and, in many cases, the techniques have not changed much over the past decades. This will change though, so you might as well get on board early and get used to adapting. If you still don't believe me, look back on evolution and how species survived ... they adapted!

The best way to assess how best to embrace change is to consider what challenges it brings and work to overcome those. The biggest challenge of change is the fear of the unknown. This fear can come in different ways: personally, someone could fear change because they like where they are, and a change could be uncomfortable or worse. From a business standpoint, there is an element of risk in change, which relates to dollars. In either situation, despite best efforts to explain the new idea, the transition plan, and what to expect along each step of the way, you must understand that at this point, it is all theoretical. To embrace change, you have to be comfortable with risk. As a side note, if you are not comfortable with risk, then get out of construction immediately. The industry revolves around risk and risk management. Yes, with risk comes the possibility of failure – that is something that must be accepted and planned for to the best of your ability – but in a greater and more positive sense, risk brings the possibility of reward. Simply put, no process, product, or technology should be introduced unless it is going to replace or improve another one. You should not implement anything just for the sake of making a change, especially if it does not fundamentally align with your personal or company values. Many people have a hard time getting past the thought of changing something that is familiar and comfortable to them. Even if you are ok with this transition and are leading a change initiative, you still need to understand how to convince others and possibly accept that some people will never change!

Leaders are change agents by nature. They know what's best for their cause and help the team get there. The key to embracing change is to plan for the risk and focus on the positive. Why are you or the organization making this change, and how it will benefit those involved? What is the main goal you are trying to achieve? You can identify a personal connection to this and the key is finding a way to convey it to your coworkers or team. It's also important to acknowledge that there will be a period of difficulty before the goal is reached. Nothing comes without sacrifice and change management is no different.

Here is an example of this looking at an estimator working on upgrading existing software. To do so requires a lot of back-end tasks such as updating/converting the database, new hardware, etc., which the technical experts and IT gurus will take care of. What you would need to focus on as a leader, change agent, or early adopter would be the employee-facing features such as a new user interface and change to existing processes. The new software will enable the team to generate more accurate estimates quicker, but it will take change. Some people might push back because they fear they will be slow to learn the new process and deemed less effective, or they just really like the current interface and do not want it to change. It is helpful to try to anticipate what issues may arise and have a response or solution ready in case it comes up either for yourself or someone else. In this example, you can give people assurance that proper training and support will be provided, sufficient time will be given to adapt, and that there will be some difficulty in the transition period but the new software will help them improve their jobs.

Change is constant in many aspects of life. By improving your ability to accept it and help others do the same, you will increase the likelihood of success and enjoyment in your career.

9

Key Skills for Success

Don't Just Network – Connect

So far, we have covered the importance of connecting to different types of individuals, who will develop your support systems and help you succeed in your career. Your network is your net worth and will be probably the single most valuable asset in your career. As covered previously, once you become proficient at your basic job skills, how you sell them and who you sell them to will determine your future. It comes down to who you know and how connected you are with them. This is where it is worth spending some time defining networking. To some, networking is schmoozing at company events, collecting as many business cards as possible, and sending out hundreds of LinkedIn invitations afterward. Sure, this may get you in front of many people quickly but the key is to leave a lasting impression on people. So, they WANT to be part of your network. Simply meeting them, flashing a smile, telling a joke, and moving onto the next victim will not have the most effective outcome. The key is creating a meaningful and long-lasting impression and eventual relationship. It is not just networking, it is connecting. Connecting takes time and deliberate action to find and strengthen commonalities and mutual benefits, which we will cover in the next section.

The construction industry is uniquely challenging because, by nature, it is highly complex and requires a large amount of collaboration between cross-functional teams and companies. The high complexity comes from

The Human Side of Construction: How to Ensure a Successful, Sustainable, and Profitable Career as an AEC Professional, Second Edition. Angelo Suntres.

the fact that each trade involved has a very technical component each requiring years of experience and training to fully understand. Even those who have been around for 30 years will tell you that they are constantly learning because there is so much involved in the individual pieces of the puzzle, plus technology and methods are always evolving. Multiply this scenario by the dozens of different trades that are required to complete a project and you can quickly see the challenges posed by the technical specificity of the industry. The other aspect that makes construction uniquely challenging is the amount of collaboration required across cross-functional teams and companies. Our industry includes so many different types of technical experts: structural steel, concrete, plumbing, heating, electrical, and ICAT, just to name a few. All these components are equally important to completing a project but are fundamentally different. Just listening to conversations within each of these circles shows that they can almost each be a separate language. However, they are all still required to come together at the perfect moment in the perfect place to complete the building or infrastructure. Wow ... that almost brought a tear to my eye, it was so beautiful. Unfortunately, it rarely works out this way in reality. There are always schedule and logistics issues and lots of communication gaps. Because all trades – and even some intercompany departments – are vastly different, silos are created, which impedes communication. Therefore, it is important to diversify your network with people and relationships that are outside of your area of expertise that can help you break down walls and interpret information.

The key principle here is that no one person can know everything. So, we must rely on others to complement our strengths and supplement our weaknesses. This is the essence of connected networks, a group – or groups – of people who can help or support each other. No one who is successful in construction knows everything; however, I guarantee that they know where to go to find the answer.

Each connection you make or sustain should focus on mutual benefit for both parties. You will get the most value from other people if you are prepared to give something in return. Some connections may not seem to provide you with that much value; the other person might have little experience or nothing like what you are looking to gain. This is the power of a network because these seemingly weak connections may link you to another person, who could be your greatest future asset that would have otherwise been inaccessible. Therefore, it is important to take deliberate action to connect with people and find commonalities, even if they seem remote. Usually, you can find something in common with someone through asking a series of questions about their experience, likes and dislikes, etc. It almost becomes a game where you continue to ask questions until you get a useful answer. Once you connect on an item, you can elaborate on that point, which can likely be the source of mutual benefit. That isn't to say that you should ask 100 questions and turn the encounter into an inquisition ... this connection does not need to be made on the first encounter. If you find that, after meeting someone, you just don't have anything in common with them, stay pleasant and keep things light and short. It is possible that more information may come to light at a separate function or meeting where it will be easier to connect if you have already established the foundation to build on.

At this point, it is worth reminding the reader that this explanation should not be taken out of context and construed to mean that you only act nice to people with the motive of having them do things for you. That could not be farther from the truth. Entering relationships seeking to add value to the other person shows care and builds trust – two key elements to a healthy, productive connection. If the relationship is one-sided, meaning that one party always takes and never gives, this might produce short-term benefits but will not provide maximum results or sustainable success and will run the risk of creating a reputation which prospects will avoid.

To summarize, networking should be viewed as more than just meeting people, shaking hands and moving on. The focus should be on connecting with people with the intent to find commonalities and add value through strong sustained relationships.

How to Do More with Less

Do you ever find that your to-do list spirals out of control to the point where you don't know where to start? Demands on employees at all levels of organizations these days are huge; there are so many emails, phone calls, and messages and somehow, we are supposed to maintain a "work-life balance." Most people struggle with prioritizing – how to determine which tasks to do first, especially when the list gets overwhelming.

For example, my brain prefers to work from the top down, tackling the oldest items (first in, first out). This makes logical sense but is not an effective way to prioritize. The key to making it all work is effective time management and there are three things to consider when planning how to accomplish your tasks: level of criticality/time sensitivity, duration of the work, and the accomplished impact.

1) Criticality/time sensitivity: Rate each item by how critical of a task it is and how time-sensitive it is. Items that are critical and time-sensitive should be prioritized while items that are not critical and not time-bound should go to the bottom of your list. All other items end up in the middle and require some judgment as to where they ultimately fall. It can be difficult to continuously change priorities and if you find that this happens multiple times a day, it could be a sign of bad management, not necessarily an issue on your end.
2) Duration of the work: When considering the priority of a certain task, a good question to ask yourself is, "How long does the task take to complete?" If the answer is "less than five minutes," do it immediately. This will help quickly shorten your list and build momentum to complete other larger, more critical tasks.
3) Accomplished impact: Another question you can ask yourself is, "What is the impact of accomplishing this task on myself/the team?" The larger the impact, the higher the priority.

In some cases, you may be able to use one of the ideas above to help sort your list. In other cases, it will be a combination. Try different things and see what works for you. Often you will find that there are important tasks that you must prioritize that you just don't want to do and other tasks that you enjoy doing that you just can't prioritize. This takes a bit of practice and discipline but is worth it in the long run. Once you start seeing results of getting more accomplished and feeling less stressed, it will motivate you to improve even more.

One of the most important tools for time management that we did not discuss in the previous section is learning how to say "no" when a task is added to your list. This can be difficult for most people, especially those with less experience who: (i) don't want to make anyone think that they are weak or difficult, and (ii) want to take on and learn a lot of new things. I am sure that you can think back to a time when someone asked you to do something, and you said yes either before considering the impact on your current workload or even knowing that you were already maxed out. It happens all the time! A very common problem is when managers lose touch with members of the team and overload people with too much work. Most people are hesitant to speak to their managers about having too much work for fear of being labelled as lazy workers or complainers. If you need to have this conversation with your leader, avoid making it personal (e.g. "you give me too much work"). Instead, explain that you are working as much as you can and are still falling behind and ask if there is anyone else available to help, even for a short period of time. Likely, they will appreciate the honesty and that you are looking out for the best interest of the company, which is getting work done properly.

The fact of the matter is that you are better off saying "no" than taking on additional tasks and failing or burning yourself out. These two scenarios will have more of a negative effect on your life than if you initially chose not to take it on and yes, there is a right way to do it. Politely refusing can be intimidating especially if you have not established a rapport with your manager or if they are unreasonable; if this is the case, then you are in a difficult situation and may need to consider consulting with a friend or coworkers if there is anything they can do to assist. Assuming that most people are dealing with adequate, reasonable managers, consider the following scenario. Imagine a worker who has a packed schedule and a full to-do list, who is probably already stressed about existing work when their manager comes into their office and asks them to take on an additional task. The worker knows that they are not able to accommodate any more work at the moment. So, the best thing to do is open up a conversation about it. You can never go wrong with effective communication. They could explain that they do not have any spare capacity at this moment and they can complete the task but it would have to take priority over other items on their agenda. They could go further and ask the manager which items to prioritize. This is an engaged approach to letting their manager know that they are at their limit without sounding lazy or complaining. By the way, if your employer doesn't trust you or respect the value you bring to the team, is that the kind of place you want to work in the first place?

Conflict Resolution – Seek Win-Win

Conflict is an essential part of problem-solving and is inherent in construction. Whether you are a project manager or a superintendent on site, you are solving problems, usually other people's problems, to add a layer of complexity. The important thing to realize is that conflict is not a bad thing and when you learn to effectively deal with it, it will improve your relationships and career. There are two common reactions when a conflict arises: avoidance and defensiveness.

Avoidance of conflict is not useful because it does not resolve anything. Have you ever avoided a task because it involved a difficult situation with someone? Or conceded in a situation against your better judgment? Yes, there will be situations where you have to learn to walk away if progress is not being made but this should not be your first action. As someone who hates conflict and is a recovering people-pleaser, I have personally struggled with avoidance for years. Only now that I am more experienced am I comfortable speaking my mind and standing up for what I believe in. I have found that even when people initially come across as defensive or combative, you can exercise patience, care, and respect and eventually they will soften to a point that you can at least have a meaningful conversation.

Defensiveness is also counterproductive. When challenged, many people feel that this is a personal attack and react by putting up a wall or being combative. If you fall into this category, you are not alone, especially in construction! I have seen some epic battles between two defensive personalities on site, especially when it comes to trade coordination and rework. No one likes to be told they are wrong, so the key here is ... effective communication! If you are on the giving end, take the time to stop and remind yourself that it is not a personal attack on you and may even be a misunderstanding. Make sure you clearly understand where the other person is coming from and try to keep calm. Similar strategies apply if you are on the receiving end when it comes to effective communication. Often when people are upset about something you said, assuming it was not intentionally malicious, they probably just did not fully understand you. So, make sure your point is explained clearly and respectfully. When things get heated, they never end well. Sure, we all need to vent sometimes but learn how to do it in a safe and harmless space.

The key takeaway here is that conflict does not go away if you scare off the other person or if you run from it. To be effective and successful, you must learn how to deal with it. It takes time and patience but will be worth it in the end.

A win-win situation is when opposing sides of an argument find a compromise so that both their objectives are accomplished with concessions. Construction, like life, is not usually a zero-sum game. There is not always a clear winner and loser and for one side to win, the other side does not necessarily have to lose. There is a middle ground that we should all strive for, whether it is trade coordination onsite, negotiating a price a bid close, or a contract after award. In every conflict, there are two sides, and each has their own idea of what success looks like to them. If you focus solely on individual factors, and both sides are firmly opposing, then there never would be a resolution. But when you introduce factors that benefit the other party, you start to seek a middle ground, which is different than negotiating strictly on one factor such as price. In the example of a price negotiation, the GC may have a budget of $100,000 that they need to meet but the lowest bidder is firm at $125,000. If both parties focus solely on price, and neither is willing to concede, it is not going to end well. If they get creative and think of other factors to sweeten the deal such as future work or alternate products, this adds benefit to the other party with a small impact on the original goal.

I am a firm believer that there is always a win-win scenario, and that we should always seek it. Let's consider the following general examples of different types of conflict outcome scenarios and the effects of each.

1) Lose-Lose:
   - Outcome: No resolution or deal is made, both sides lose
   - Impact: Benefits no one, likely will impact business in the future for both parties
   - Not sustainable and hurts relationships
2) Win-Lose:
   - Outcome: One side gets what they want and the other concedes totally
   - Impact: Benefits one side only, the other will probably avoid similar situations in the future
   - Not sustainable and hurts relationships

3) Win-Win:
   - Outcome: Both sides come to an agreement with mutual concessions
   - Impact: Benefits both sides, likely to foster or maintain business continuity
   - Mutual value-added and sustainable relationships built

In some cases, it is better to consider a lose-lose if you are facing the losing side of a win-lose, and similarly to saying no to additional work discussed in the previous section, there is a respectful way of doing this.

Always seek win-win. You will not be surprised to hear that the conclusion to this section is that the ability to seek this type of resolution relies on your relationships/effective communication skills and is built on mutual respect and trust. Without these elements, reaching a win-win will be difficult.

10

Provide Solutions to Big Problems

Design Problem

In the construction industry, there are many existing gaps that create problems in how the industry operates. Some have been around for years and continue to grow, while others are caused by recent changes in the workforce and advancement of technology. Regardless of the cause, these gaps have a direct impact on every company today and need to be addressed if we want to have a successful future for the industry. In this section, we will identify the three critical issues that are present in all organizations and steps we can all take to narrow the gap and minimize the impact of each. It is important to understand how all of these affect your team and what you can do to help solve these big problems to improve your organization and boost your career.

The first type that we will discuss is the gap between design and construction. On any type of delivery model, whether it is design/bid/build, design-build, IPD, traditional or collaborative, there is a point when the consultant team of architects and engineers completes the design. Unfortunately, the design will, inevitably, contain gaps when compared to finished construction. These gaps can go unnoticed for a very long time and cause significant schedule and cost impacts to a project. I am not saying that it is an issue with the knowledge or credibility of designers; there are so many fine technical details that are required to complete a project that it is not reasonable to assume that the design team will capture it all. Nor is it the responsibility of the builder to ensure that they have

The Human Side of Construction: How to Ensure a Successful, Sustainable, and Profitable Career as an AEC Professional, Second Edition. Angelo Suntres.

every single design detail covered. This is easier to manage in newer collaborative delivery models where the design team is part of the construction consortium, but it is still present and must be managed closely to minimize risk of schedule and cost impact, which we will cover in the next section.

In traditional design-bid-build projects, this risk is solely on the owner as the contractor will typically submit their bid based on plans and specifications with any missing components charged as an extra or change order. In either case, this creates unknowns when it comes to the overall budget and schedule of the project from the owner's perspective. There are pros and cons to different delivery models including shared risk, speed to market, and overall cost certainty that must be considered when deciding which route to take. This book does not get into the contractual details of each project delivery format, but it is recommended that you become familiar with all models as you never know what your future has in store and how it will impact the gap between design and construction on your project.

Before you can attempt to solve a problem, you must acknowledge that it exists. Doing so requires some level of introspection and realization that no one person or team is perfect, and they cannot be expected to know everything. We are all part of the problem until we begin to be the solution. Everyone has their scope of expertise and zone of genius. Part of this is knowing the boundary where yours ends and another's begins. Addressing the gap between design and construction relies on the relationship between the design team and the build team and starts with mutual trust and respect. Countless times I have heard builders criticize designers saying that they have no idea how to build, despite years of schooling. On the other side of the table, you have the architects and engineers, who criticize builders for having no idea how to design, despite years of experience. The truth is that both sides are correct, but this needs to be viewed as a unifying rather than a divisive topic.

The fact is that there is a reason that designers design and builders build; the industry is too complex for one person or company to do both. The sooner you realize and acknowledge that you don't know everything, and you do need others to help you, the sooner this gap will begin to narrow and the faster innovation and collaboration will increase producing positive results for your project and organization. It is ok to say that you don't know and to ask for help and if you have cultivated the types of relationships that are based on mutual trust and respect, you will all feel safe to do so. The real magic happens in this state of vulnerability, knowing that someone else has your back and vice versa.

Another barrier that could be attributed to the increase of this gap is the reliance on technology and its impact on connection between people. Too often we rely on emails and remote meetings where a lot of communication is lost. In my opinion, nothing beats face-to-face communication when dealing with a troubleshooting or brainstorming session where a high level of collaboration is needed.

The best experience comes from making mistakes. So, make sure to lean on more experienced colleagues so you don't repeat the same errors they did. And remember it is ok to take occasional risks ... just be smart and don't lose too much money. Sometimes larger organizations struggle to communicate lessons learned concerning both successes and failures, causing the same issues to happen on multiple sites. This is another great example of why connecting and expanding your network is so important, as we discussed earlier, to have more contacts to reach out to and learn from.

Worker Problem

The worker problem refers to gap between people in the field and in the office. This can exist between the site supervision team and the project support staff (Supers/PMs) but is more drastic in the on-site versus head office roles for similar reasons to those mentioned above when we discussed the Design/Construction gap. The main cause that I have noticed comes from two main sources: education/experience and work schedules.

Field staff traditionally are likely to have more experience than education, while office staff are more likely to come from a post-secondary or master's program. This is not to say that all field staff are not educated or are less intelligent because they may not have attended university, nor does it mean that all office staff are intellectually superior, but rather this is a summary of observations that I have made throughout the course of my career. At the end of the day, I personally do not think education has much of a bearing on where you ultimately end up in life. Schooling is great for teaching technical principles and theory and can help open doors where there are minimum job requirements, but what matters most is how you learn to apply these skills in real-life applications.

Another point of contention that I have noticed over the years is that field staff usually work different hours than office staff. Site work typically involves early mornings or late nights (sometimes both) whereas office roles have a bit more predictability when it comes to hours of work. Both roles can have a more demanding schedule than other industries, which is a result of the nature of construction. If the team is facing an issue or large contractual milestone, it will require extra effort from all members to achieve the goal. Other times there will be more flexibility with your work schedule. It is give and take.

Your experiences may not align exactly with the examples above; however, I am sure that you can think of similar examples of conflict that arise out of differences in job conditions between site and office. Maybe I am too optimistic and try too hard to find a "silver lining," but I see this as an opportunity and example of why the construction industry is so great. There are so many different facets of the industry that allow different types of personalities and people to find their niche and excel. We all just need to find a way to get along.

By this point, you can probably guess what the solution to this gap is, but I'll repeat it again just to drive the point home. Establishing a healthy relationship based on mutual respect and effective communication will help address this gap with the understanding that we all have a role to play in the bigger picture of the project. Just like my suggestions for narrowing the design/construction gap, both sides here bring forth not only their job functions but experiences, strengths, and weaknesses. An effective team will be able to integrate the two sides and find ways to complement and supplement each other's skills. Acknowledging that there are differences is the first step to reconciling them in any situation. When dealing with the example of Superintendents/PMs, we can all agree that these are very different but equally important positions on any project. You cannot and should not compare the two based on the technical training or hours worked because the requirements are fundamentally different. The important thing here is to look at the big picture and how each person contributes to the main objective of the project. Notice how this principle aligns with the "finding commonalities" theme that is critical to relationship building.

Effective communication can go a long way to help discuss the differences that both roles have and how they complement each other. Again using the Super/PM example, the site team cannot install equipment that was not procured by the PM team. Conversely, the PM team cannot bill/get paid for work that the site team has not completed. It is important to realize where your expertise and responsibility end and the other party's begin in order to keep the project progressing. To illustrate, picture a seesaw that you played with as a child. Each person on opposing ends takes their turn pushing off the ground to keep the activity going, but if there is too much weight on one side or the person does not push off the ground hard enough, the game is over, and the kids get upset! The responsibility for projects is always ebbing and flowing between the site and office. So, both sides need to maintain respect and communication to get it done!

Generational Problem

Probably, the most prevalent and impactful of all the gaps discussed in this book is the one between generations in the construction industry. Ever since I started working in 2007, we have been talking about the labor shortage that would eventually occur when people of the baby boomer generation started to retire. Even though it has been the subject of much conversation and debate, not much was done to plan for it and, so, here in 2024, we find ourselves in this exact situation. In all organizations I have worked, I have noticed that the demographics are similar: you have a lot of workers at or near retirement age with a wealth of experience and knowledge, a small number of people (usually in middle management) in the 40–50 age range, and then a large percentage of younger employees with little experience but a lot of education. This is not to say that this is reflective of all organizations, but this is what I have observed. There appears to be some sort of barrier that impedes the extraction and transfer of all the experience and knowledge from the more senior employees to the younger ones. I believe that this can be attributed to a few differences in work and lifestyles between the two, namely motivation and communication. Yes, there are other factors that impact these items like how long they have been in the industry, background, etc., but for the purposes of this book, we will focus on these two.

The motivational differences between these two generations can be summed up in one statement: older generations worked to provide basic necessities for themselves and their families while younger generations are more motivated by self-driven means like work–life balance and personal enjoyment/satisfaction. This is not to say that one side is right or better than the other, and, of course, this is a generalization, but I believe that this creates a barrier for the two sides to connect.

There are also communication differences that pose issues. Younger generations prefer to communicate via text messages, emails, and remote meetings while older generations favor phone calls and in-person get-togethers. This may seem like a simple one, but I believe it has a profound impact. When you are not able to connect in your preferred method of communication, a lot of what you have to say will be lost. Conversely, if you are on the receiving end of something that is conveyed through an outlet that is not your preference, you will have a more difficult time understanding what the other person is trying to communicate.

In either situation, the most important part – information transfer – is negatively affected, which impacts the future of the industry. When the more experienced team members retire, they will take all their knowledge with them. It would be a shame if we let all of that information and lessons learned go to waste because we could not communicate effectively.

The key to closing this gap is to again acknowledge that there are issues that need to be addressed. The issue of motivational differences is one that needs to be addressed by HR or senior management as they may need to change the way their employees are rewarded or expected to work. For example, there has been recent pressure on employers to provide a remote work option; however, this does not fit in some organizations or with employees of a different generation. Creating awareness and information on differences like these including where they stem from and why could help both sides to better understand where the other is coming from and reduce frustration between team members. The responsibility to address the issue of communication differences falls on the team leaders as each specific project or team will have its own requirements based on the types of people involved. A potential solution here is to draft a communication style matrix indicating what types of exchanges should happen via which mediums to avoid a one-size-fits-all approach. An example of this could include:

1) Any issue that requires immediate explanation (e.g. a leak in a certain area of the building or a power failure) to be communicated via phone call to your supervisor; or
2) Confirmation that the deliveries have arrived to be communicated via email to Joe

Some workers may need support or encouragement to use different types of communication methods (e.g. training on how to write an effective email, benefits of calling versus texting, taking time to show a coworker how to use an app, etc.).

The generational gap in the construction industry is not just a problem, it can also be viewed as an opportunity. It's a chance to bridge the divide and create a more cohesive, productive, and innovative workforce. The key lies in understanding, respect, and adaptation for and with others, not staying with status quo of "this is how we've always done it" or "it was like this for me, so get used to it."

Respect is equally important. Each generation brings something unique to the table. The older generation has years of experience, practical knowledge, and wisdom. They have been through numerous projects and have valuable insights that can guide decision-making and problem-solving. On the other hand, the younger generation is tech-savvy, adaptable, and innovative. They bring fresh ideas, are quick to learn new technologies, and can drive the digital transformation of the industry. Recognizing and respecting these strengths can foster mutual respect. But how do you break down the barriers preventing this recognition and respect? The answer lies in true understanding. Both generations need to understand each other's motivations, communication styles, and work ethics. This can be achieved through open dialogues, workshops, and team-building activities. At first glance, these platforms can come off as cheesy, forced, and unproductive but, if done properly, they can serve as a venue for each generation to share their experiences, perspectives, and expectations. It's about creating an environment where everyone feels heard and valued.

Adaptation is another piece of the puzzle. The construction industry, like any other, needs to adapt to the changing times. This includes adapting to the changing workforce, their needs, and their wants. Companies need to rethink their strategies, from recruitment and retention to training and development and have a serious look at their accepted behaviors – cultures. They need to find ways to attract and retain younger workers, such as offering competitive salaries, career advancement opportunities, and a positive work culture but they also need to invest in training programs, not just for the younger workers to learn the ropes, but also for the older workers to keep up with the latest technologies and practices.

Moreover, companies need to address the communication gap. This could mean adopting new communication tools that cater to the younger generation's preference for digital communication, while also maintaining traditional communication methods for those who prefer them. It's about finding a balance and ensuring that everyone is on the same page.

In conclusion, the generational gap in the construction industry is a complex issue that requires a multifaceted approach. It's about building bridges and breaking down walls. It's about fostering a culture of understanding, respect, and adaptation. And most importantly, it's about recognizing that every generation has something valuable to offer and that together they can build a stronger, more resilient industry. As we move forward, let's remember that it's not just about filling the labor shortage, but also about harnessing the full potential of our multigenerational workforce. After all, the future of the construction industry depends on it.

There are likely many other gaps in the construction industry, but these are three key areas that we all need to be mindful of if we want to be successful in the future. Being a frontrunner in identifying and helping solve these problems in your organizations will surely be an effective way to distinguish yourself as a leader and help solve the industry. It requires group effort. So, we need all hands on deck.

11

The Special Sauce

Know the Basics Really Well

This might sound bad, but you do not have to be an expert at your job to excel at it, make connections, and get promoted. You just need to know how to do the basics, really well. This might sound simple, but it is not easy and it does take time, patience and hard work. In my personal experience, I find that it takes 8–10 years in your career to really get a grasp on the basic technical skills required to understand and be a productive member of the construction industry, whether it is project management, estimating, site supervision, or other roles. This may seem like a long time but there is so much to learn and so many topics to cover, which is why it is important to have realistic expectations of how long it will take you to develop the skills and attributes required to make the next step in your journey.

This book emphasizes the importance of people skills and building connections, but you cannot neglect the fact that you have basic job duties that you need to complete. You could be the friendliest, best talker, and most influential person on your team but if you're not completing your day-to-day tasks, you're not going to be employed for very long. Spend some time to get to know how your organization operates and what skills you need to focus on in your role to have the most impact and generate the best results. Additionally, take some time to get to know your coworkers and how their roles fit into the larger picture of

The Human Side of Construction: How to Ensure a Successful, Sustainable, and Profitable Career as an AEC Professional, Second Edition. Angelo Suntres.

the organization. Ask what they do on a day-to-day basis, what software or tools they use, and challenges that they face. If done properly and genuinely, it will show who you are, and widen your knowledge base and experience, which, at some point, you will be able to showcase to your manager or other members of the leadership team and help distinguish yourself from the others to set you up for more responsibility or advancement. Don't get discouraged if there are certain basic skills that you find are not your strength and remember that it is okay to ask for help. You should not be expected to know everything and it is better to ask for help than to struggle and fail.

As you spend time finding what you like doing, mastering the basic skills, and making connections, you should develop a clear picture of where you want to take your career. You will reach a point where you need to decide between two different options. The first is growing within your role and becoming a technical expert and the second is growing out of your role and expanding your responsibility into more general management-type roles. There is no right or wrong answer and it is entirely up to you how to measure success in your career.

This section of the book is tailored to those who do not wish to remain in the technical field and want to take more steps into management or senior leadership. It is worth mentioning again that there is nothing wrong with deciding that you love doing what you are doing and that you do not wish to take on the additional responsibility or scope of a management role. Each person's journey is individual, so only you can decide that.

Depending on the position you are in and the position you would like to eventually get to, the method and steps to get you there will vary. As previously mentioned, I believe that to progress to the next step in your career, you really need to master the one you are currently in – both the technical and the people side – or else you will struggle at the higher level. For example, if you currently hold a Project Coordinator role and you wish one day to be a director or VP-level executive, you cannot expect that by starting to act as a director or VP will automatically qualify you for that role. You need to gain the experience and exposure required to build up to that.

The key to any plan is having a clear goal with a realistic timeline. Otherwise, you are setting yourself up for disappointment or failure. You must carefully assess your current job function and your relationship with your manager, including the type of leader they are, to find the right mix of technical and people skills that you need to develop, grow, and advance. From my experience, those who express more ability at the technical end of the spectrum tend to be viewed as a critical resource in that role, which translates to an impediment that may hold you back, especially if you don't have the strongest leader. In this case, success of the team may depend on you staying at your current level. Since construction is a very technical industry, there are lots of this type of personality and without them, the industry would fail (again, everyone has their role and parts of play). If your background, experience, or personality puts you more on the technical side of the spectrum and you feel like it might be limiting your career, it is never too late to change, it just may take some extra time and effort to learn other skills discussed in his book to develop your interpersonal communication and connecting skills.

Focus on People, Not Things

Since relationships will ultimately determine your success in life, it is important to at least have some sense of interpersonal skills and to be an effective communicator. Even if you choose to stay in a strictly technical role, chances are you will still have to communicate with others on a daily basis. So, you can't deny the importance of people skills. As previously discussed, too much focus on the technical side of things may pigeonhole you into a certain role. Conversely, focusing too much on people skills opens a different set of challenges where you may lose the edge on effectively completing day-to-day tasks, so the key is identifying when and where to apply both skills.

For example, if your manager called you into their office and asks you a direct question about a serious matter such as a recent major cost overrun or scheduled delay, it is probably not the best opportunity to discuss what they were up to over the weekend or how their kids are doing. Make sure to read the room in each situation and answer any questions directly. If there is a pause or natural moment for a subject change, that is your opportunity to get into more personal discussion. This does not come easily to most people and will take some time to learn as part of the art of conversation. If you are already in a management role, then much of your day will be spent focusing on people. So, it is more appropriate to talk about general topics such as how things are going, what challenges a colleague is facing, and productivity or other team dynamics.

Instead of waiting for these opportunities to exercise your people skills and connect with others, you can actively seek them out. A perfect example of this is coincidental meetings at the coffee machine or the infamous water cooler. The key here again is reading the room and understanding what others are looking to get out of a conversation. If they are providing short answers and their mannerisms indicate that they are just trying to get back to their desk, it is not the best time to get into a long conversation; so keep it short and pleasant. Conversely, you do not want to develop a reputation for being too chatty and taking up others' work time. So, again assess the situation and how it would be perceived by others including your management team.

One of the most common mistakes people make in their careers is not leaving relationships on good terms. This type of situation most commonly happens when an employee quits or is terminated but can also apply to coworkers, friends, and associates. The enjoyment that you may feel from burning a bridge is not likely worth the impact it may have on your future. When dealing with networks and your connections that form your support systems, it can sometimes be seemingly weak or insignificant connections that put you one step closer to somebody or something that you need access to. If there is one thing I have learned in my years in construction, it is that even in a large geographical area, it is a very small industry and word travels fast. You also never know who is going to be your next boss.

The key principles and recurring themes of this book remind us that human connection and relationships are at the core of your success and that they are based on the elements of mutual trust, respect, and care. A great indicator of someone's character is how they react to someone who has treated them poorly and can be an opportunity for you to exhibit intelligence and self-control by taking the high road. There may be cases where others severely mistreated you and you would get loads of satisfaction from telling them what you really think and putting them in their place. Even if you are in the right and this might feel good in the short term, it is not likely to help you in your long-term success. Emails can be recalled; words cannot.

12

Smash Limiting Beliefs

Failure Is Not the End

How often have you considered making a decision or taking on a new job only to talk yourself out of it because you think you can't handle it? I am sure you can think of a time when you doubted your own abilities or worried that the risk of failure was too great. Looking back, how many of those situations do you regret not taking action on? I have taken many risks ranging from new tasks that I was unfamiliar with to new positions that were outside of expertise at the time and yes, it was scary but wow, was it ever worth it! Of course, there are also some decisions that I would have made differently had I known then what I know now but learning is part of the journey when you try new things.

If there is one piece of advice that I would give my former self when just starting out in the industry, it would be to find the people and place where you feel safe to make mistakes. There is no better learning experience than trying it for yourself, and failures make for the best lessons. Of course, your level of risk should be relative to the amount of knowledge and experience you have in the field you are in but there is always a chance of failure that you must embrace to try something new. That's where the magic happens!

Chances are you have some tolerance for risk, or else you would not be in the construction industry. In my personal experience, the best lessons I have learned in life both about my career and myself have come from

The Human Side of Construction: How to Ensure a Successful, Sustainable, and Profitable Career as an AEC Professional, Second Edition. Angelo Suntres.

stepping outside my comfort zone, taking calculated risks and inevitably failing. Upon reflection, I have found that those failures were actually little successes in disguise.

Again, I stress the importance of finding a safe environment where you have the support and approval you need to make decisions and potential mistakes. It is always prudent to inform your manager of steps you are taking and the anticipated outcomes including the risks associated with the actions. It is important to evaluate the environment you work in before you make any decisions and consider the impact carefully to not negatively impact your career.

This may be very difficult for some people who are not comfortable with change or trying things outside of their comfort zone. This is usually a personality trait but if you feel that your current risk tolerance is limiting you, it is possible to expand it starting small and growing in the size and complexity of the chances you take over time.

The most growth happens when you embrace change and try new things. I am not encouraging anyone to make irrational or irresponsible career choices, but you must make the best decision you can with the information available at the time and have faith that the unknown will work out in the future.

Failure is simply the end of one thing and the beginning of another. It is okay to fail and there is a right way to do it. To fail, you must try, which takes courage and faith. So, the fact that you made the attempt at all deserves a level of congratulations. To try means that you have embraced the possibility of failure and you did it anyway. Our brains have a way of making perceived threats (failures) seem worse than they actually are to protect us from getting hurt. Usually, the actual result of the failure pales in comparison to how you perceived it. I am sure that we can all recall a time when we experienced failure and thought that we'd never recover or that our lives would be drastically impacted forever, only to find later that it is not much more than a memory.

Now think back to a moment in your life where you tried something new, despite hesitation or nervousness. A perfect example is learning how to ride a bike. Whether this happened as a child or an adult, learning how to ride a bicycle can be intimidating. There is speed involved on only two wheels and your balance is the only thing protecting you from crashing to the concrete! Sure, it will hurt if you fall, and you could get injured, but most people try anyway. Most people do fall multiple times before being able to bike correctly but I am sure we can all remember this moment in our own childhood or our children's experiences, and we all share the same joy and excitement the first time we rode alone. This example may seem simplistic since it does not involve the complexities of relationships and careers; however, the same principles apply when compared to another scenario. Imagine that you are an estimator and you are asked to take the lead on a new pursuit for the first time and you are worried that you will struggle. The following table illustrates the common points:

Description	Bike	Estimate
Objective which requires something new	Riding a bike	Leading a pursuit
Potential for failure	Falling down	Unsuccessful pursuit/not closing
Impact of failure	Physical pain, embarrassment	Mental pain, embarrassment
Support system	Parents, other kids	Manager, coworkers
Reward for completion	New skill learned, freedom to ride	New skill learned, more responsibility/promotion

This is a good exercise to do to compare risk versus reward in any scenario you face.

The key takeaway is that failing is common and is part of your growth; it is not the end of the world even if you think it may be and – when you have many healthy relationships and connections – despite those that may make you feel belittled, ashamed, or embarrassed, there will always be people around to guide you away from failure and to help you recover when it happens. Plus, the people who do not support you likely have their own failures that they have not properly resolved with themselves.

Mistakes Make You Better

The most memorable lessons in life are learned from making mistakes because they are associated in our minds with pain. The key is to make small, calculated mistakes which will only hurt a little bit and not drastically impact your relationships or career. Learning what not to do on your own is more effective than having someone tell you what to do because the former involves figuring it out yourself. This process can be extremely powerful in your development and how you are viewed by others. In construction, the people who get things done are the ones who get the most respect, recognition, and success. So, be a doer, even if it means making mistakes. Organizations should provide an environment that allows employees to make mistakes without fear of punishment if it was thought out and reasonable. Everyone should be allowed to make mistakes, but don't make the same mistake twice!

The act of doing and failing is a great way to build resilience and character and is the best experience, especially in the construction industry where every project brings its own set of unique challenges. Construction is rarely a one-size-fits-all industry and what worked on a previous project or estimate may not apply to the next. That's why it is important to get comfortable making small mistakes that don't cost much time or money. Most times, the process of trial and error, if done correctly, will result in less time and money spent than overanalyzing before acting. I can think back to multiple times in which the meetings, planning, and coordination of certain activities took longer than if we tried and failed twice.

Another important point to understand is that everyone, and I do mean everyone, makes mistakes. Yes, even people who seem like they have everything together and come off as perfect. Making mistakes does not make you less intelligent or worse at your job than others, unless you keep making the same mistake ... that may be cause for concern. Failing, or making mistakes, is not the end; it is up to you how you react to the mistake you made and what you do with the information and experience that you have garnished.

Many people are ashamed or embarrassed about their mistakes for fear that they will be punished or lose the respect of others. Unfortunately, in the construction industry, it is not uncommon to be chastised for making a mistake. So, it is not surprising that many feel this way. This is an outdated stereotype that we all need to combat in order to make the industry more welcoming to others. The world is just different these days and what once was thought of as harmless joking or teasing is actually harmful to individuals and damaging to the industry.

Going back to the main elements of a healthy, sustainable relationship – trust and mutual respect – the more you can form a strong connection with someone, the more likely they are to open up to you and be vulnerable, which includes sharing mistakes. Before anyone can share any information about themselves as personal as a mistake or failure, you need to establish that connection and develop a rapport. I like to say that the more you trust, the more you'll share, which is what I have observed through people I have connected with in my life. Organizations should strive to create caring environments and provide psychological safety, which allows people to be comfortable being themselves without fear of criticism or judgment. This is how real connections are made and where people are most willing to share their true successes and failures. This topic is especially important given the generational gap previously discussed where we all need to transfer as much knowledge as possible from more experienced workers before they retire.

The lessons learned from your own failures are singular, meaning that you must experience them yourself to learn. However, if you open yourself up to connecting with others and allow them to share their experiences, all knowledge is multiplied. It frustrates me to no end when teams within the same organization make the same mistakes repeatedly on different projects resulting in a huge waste of time and money. You can solve this problem for yourself by making the right connections with the right people including similar roles on other projects, teams, or companies. If some mechanism to transfer knowledge across your organization does not exist, it may be an opportunity for you to bring this up with your manager and suggest ideas or ways to implement it. Showing some initiative never hurts when communicated properly.

This topic is actually a very useful tool that can be used to expand your network and make meaningful connections. Ask around your office or project if anyone has worked on items similar to that, which you may be starting or having difficulty with. After receiving some contacts, you can reach out to them and request a discussion regarding their experience with the project/problem you are working on including challenges and successes. Make sure to mention your coworkers' names for commonality and to increase the probability that they will return your message or call. If you don't ask, you usually don't receive.

The key point here is to remember to build trust and mutual respect with someone for them to feel that you care enough for them to be vulnerable and spend their time with you. The more we help and learn from others, the stronger the team gets.

13

Big Picture Mindset

Keep the End in Mind

Whether you are planning a family vacation or building a project, there is an end goal in mind. Before you start out on an adventure, it is important to get a clear picture of what the objective is in very specific terms that will serve as the guiding light or final objective for the team to meet. This vision is usually determined by senior-level management when related to organizational efforts, or senior site staff when related to a project. Without a clear vision, any work that involves collaboration between people or teams will go to waste. Communication of this vision is also critical as well. The key is to get everybody on the same page and moving in the right direction. This is especially important in an industry like construction where projects could take years and involve hundreds or even thousands of people to build. Imagine if every one of these people had a different view or a vision of how the end product would look ... it would be chaos.

This may seem obvious on a construction project where there are plans and specifications that show exactly what the building will look like and how to build it, each step of the way. However, because construction involves so many different technical experts and different skill sets, it is easy to find yourself or your team working with your head down, potentially losing sight of the end goal. These "blinders" are common and need to be considered by all leaders to keep teams focused

The Human Side of Construction: How to Ensure a Successful, Sustainable, and Profitable Career as an AEC Professional, Second Edition. Angelo Suntres.

and effective. What also tends to happen is that one team or trade can get so involved in their own work that they lose sight of the fact that coordination and collaboration with other people are essential. There have been times where one trade has proceeded with their own work without considering other groups' installations or schedule logistics and caused added interferences, schedule delays, and cost overruns. While each company is usually out to increase their own productivity and profit, no one can neglect the overall team effort that's required to complete a project, not to mention the cost and schedule impacts of rework!

Keeping the end in mind is an ongoing continuous effort; you cannot just "set it and forget it." It also helps everyone focus on the big picture, which can help with team motivation and morale. My favorite story regarding this topic came from a talk delivered by Simon Sinek, an expert in human performance, which I will paraphrase here.

> One day, a man walked past a construction site where two masons were laying bricks. He stopped and asked the first mason if he enjoyed his job to which he replied, "I hate my job! I am out here in the hot sun, I lay brick after brick, the days are long and it feels like I'll never be done." The man thanked the first mason for his time and walked down the street and asked the second mason the same question to which she responded, "I love my job! Yes, the days are long and it is hot and I am here laying brick after brick doing the same thing repeatedly, but I am building a cathedral!"

This story illustrates the power of understanding how your daily tasks contribute to the end goal, which will provide a sense of pride, accomplishment, and motivation for the team.

My grandfather used to tell me if you walk with your head in the clouds, you will trip over the objects in front of you, but if you walk with your head down, you will lose sight of where you are going. The sweet spot is finding the happy medium to stay on course and reach your final objective while appreciating the scenery and experiences along the way.

The ability to see and focus on the big picture is a must in the conceptual and planning phases of a project or career but this alone will not help you accomplish your goals. In order to reach the objective, you have to understand the steps to get there, and the resources required to execute – who, what, when, where, and how. To help illustrate the balance between big-picture thinking and understanding collective individual efforts. imagine working on a jigsaw puzzle. The main image is clear on the box but there are thousands of pieces that you must fit together in their own unique position and orientation to make it all come together. In a management role, it is your responsibility to make sure that all the pieces fit on your team, that there is alignment with everyone, and that everyone is working toward the common goal. If you are part of a team, it is important to understand how your piece fits with your immediate surroundings and the role you play to achieve overall success.

The company's vision needs to be clearly communicated to every person in the organization. Senior leaders are excellent at this but, in many cases, communication breaks down at the middle management level like a game of "telephone." When workers on the ground level who are executing the plan do not clearly understand the ultimate vision, chaos can ensue. This can lead to frustration and turnover since workers sense some misalignment between what senior management has communicated and directions they are receiving from their immediate managers.

If workers understand and share the company's vision but do not know how they fit into the grand scheme of things, they will struggle to find their place and will not be able to reach their full potential. Therefore, as leaders, it is so important to connect with others to provide ongoing guidance and support as required to help the individuals succeed as roles, responsibilities, and people change over time. As elsewhere in this book, when I use the term "leaders," I am not referring to management or any particular role or title. I believe that we are all leaders in our own right and have a duty to ourselves and those around us to inspire everybody to realize their potential, find their place on the team, and be the best that they can be.

Don't Sweat the Small Stuff

Your time is your most valuable resource; so, use it wisely. Any time and energy that you spend worrying about little things takes time away from accomplishing big things. By little things in this sense, I mean activities, thoughts, or worries that do not bring any value and do not contribute to the overall success of your project or career including tasks on the bottom of your priority list, worrying about things or situations that are totally out of your control, or dwelling in negative or unpleasant places.

Ultimately, you cannot control many situations in your life, namely how others act. So, dwelling on past situations adds no value and can really bring you down. This is not to say that you should not properly deal with past experiences such as traumatic events; these actually need to be properly processed to avoid future issues. I point this out because it has affected me in my past several times and each time I spent way too much time in a place of worry or anger over a past event that I could do nothing about. Try to focus on things within your circle of influence that you can have a direct impact on fixing or improving.

As an example, if you are a Project Manager on a construction site and you disagree with a decision that senior management has made for your project, you are well within your rights to voice your concerns tactfully and thoughtfully as the decision likely impacts you professionally and maybe even personally. That said, if the decision has already been made, it is ultimately out of your control and unlikely to be reversed. At this point, you have two options: you can accept that the decision is final and do your best to deal with the outcome keeping in mind that you've already voiced your concern, or you can continue to oppose it. Choosing the latter will usually result in tension, anger, frustration, and awkwardness for anyone involved. Sometimes, you need to learn when to drop an issue before it becomes a roadblock for your career. If you find yourself struggling with situations like this, either on the giving or receiving end, it is never too late to do some critical thinking and talk it over with a trusted friend or colleague before deciding for yourself.

The important takeaway here is to use your time and energy effectively; you need to differentiate and focus on main issues that you have control over and learn how to live with the little ones that you can do nothing about.

Keeping a positive attitude in the face of adversity has a huge positive impact on people around you and your collective team. It also helps keep people on track by focusing on solutions instead of being consumed by the problems they face. This is especially important in the construction industry because we are in the business of solving problems. Day in and day out, there are constant issues that require our attention and sometimes it is difficult not to fall into a negative state of mind. Everyone has bad days and is susceptible to negativity, but try to avoid it at all costs.

Generally, I am a very positive person and even I need to constantly remind myself to stay positive and lift other people up even in difficult situations. The important thing here is balance. Keeping a positive mindset is only useful if you can, at the same time, confront the obstacles and difficult facts that you are facing. Taken to the extreme, positivity can become toxic if it disregards the challenges and difficulties people are experiencing at the moment. An example of this would be if you were in a project and facing an equipment delay, that would impact a huge milestone and trigger financial penalties and you said, "That's okay! At least the last equipment delivery came in on time and we're all doing a great job and we will finish eventually!" This type of comment, though well-intended, disregards the impact and downplays the severity of the consequences that the team is facing and may be looked upon as irresponsible. A better way to handle this situation would be to acknowledge the difficulty and challenges that the delays will cause the team and the project and do everything you can to improve the current situation or make up for it in some other way later on in the project.

To summarize: avoid negativity, be positive, and lift others up, but do not be so blindly positive that you dismiss obvious problems with major implications.

14

Fair Treatment for All

Diversity, Equity, and Inclusion

The topic of diversity, equity, and inclusion (DEI) has attracted a lot of attention recently in the architecture, design, and construction industry. We have probably all experienced some form of inequity in our life or career, from pay inequality to racialized insults and so many other examples. We must stop all forms of intolerance for others regardless of their background, skin colour, sexual orientation, etc. This book does not delve into the politically charged acronym DEI, but rather it is about what the principles stand for – leadership and connecting with others regardless of the differences they may have and the least we can do for anyone is to treat them as equals. Remember, respect is a key component of building the foundations for a strong relationship.

Recent movements for DEI have taken on many forms but can include increasing the number of diverse populations in all levels of organizations from senior management roles to entry-level positions, fostering mentorship, and other types of important relationships with a focus on individuals from underrepresented groups, and even training opportunities. These types of efforts will ensure that more diverse groups of people are implemented into the workforce, but this alone will not stop the problems we face. Simply bringing more underrepresented groups into construction will fail if those people are not properly supported with the right environment. Hiring quotas and mandatory training do not equal diversity and inclusion. We must do more to improve the

The Human Side of Construction: How to Ensure a Successful, Sustainable, and Profitable Career as an AEC Professional, Second Edition. Angelo Suntres.

culture of our organizations to ensure that everyone has the support and tools they need to succeed. This goes beyond physical tools and training; we must affect change in the broader environment and culture.

In the construction industry, we especially need a concerted effort to promote these concepts to attract a more diverse range of people and make the industry more welcoming for all. There have been challenges with promoting the industry to two groups of people in particular: women, and Black, Indigenous, and People of Colour (BIPOC). Traditionally, construction has been dominated by white males; over time has perpetuated the stereotype that women and people of the BIPOC community are not welcome. I believe that we have a long way to go, but steps are being taken and seeds are being planted now to make the industry a better place in the near future.

Speaking candidly, this is an uncomfortable topic for many because it requires us all they think long and hard about the way the industry and everyone in it operates and requires facing some difficult truths. Without confronting and pushing through this personal discomfort, the collective industry will not improve. So, we have a duty to do the work, regardless of discomfort, and make it right.

With this movement toward DEI in the construction industry come some myths, misconceptions, and misunderstandings. The term "affirmative action" is defined as the practice or policy of favoring individuals belonging to groups regarded as disadvantaged or subject to discrimination. Many people feel that this is unfair and that people should be selected for positions and promotions based on merit alone. In a perfect world, I agree with this last statement, but the truth is the real world is far from perfect and the construction industry is no different.

Many factors including unconscious bias and availability bias have negatively impacted and severely limited a lot of people from underrepresented groups in the construction industry. Combating unconscious bias requires a deep level of reflection as to why we choose certain people over others. Ultimately, as humans, we have an innate need to protect our survival and surround ourselves with those who we are comfortable with, i.e. people who look, talk, and act the same way as us. Since construction has traditionally been dominated by white males, it is no surprise that we find ourselves in the situation we are in.

Another misconception about this movement is that most people will now be negatively impacted as companies increase the level of diversity within their organizations. This objection comes from a scarcity mindset and is not going to help anyone. Regardless of who is receiving the job, promotion, or bonus, one person getting more opportunities or responsibilities does not mean that they are taking opportunities away from anyone else. Remember that construction, like life, is not a zero-sum game and by supporting and caring for others, we will make the industry a better place for all. Excel in your own way and you will get recognized for who you are, what you do, and how you treat others. At the end of the day, that is what really matters to your success and happiness.

A rising tide lifts all boats.

Summarizing, it is important to understand what DEI is and, more importantly, what it is not. I hope this is clearer for you now; however, if you would like to learn more, ask your HR/People and Culture department, or do your own research online. It will be worth it for you and everyone else!

How to Change the Industry

The main factor driving the movement for diverse and equitable workplaces is ensuring fair treatment for all, but it has the additional potential to solve another current issue that is plaguing the construction industry: the labor shortage. We are currently feeling the effects of the labor shortage at all levels of organizations in the construction industry from entry-level employees to senior-level executives. Simply put, the rate at which the industry needs workers at all levels is increasing faster than there are people to fill the roles. This is attributed in part to the sheer volume of project work required to support population growth in the infrastructure, residential, and healthcare sectors. High immigration rates and an aging population are the main drivers behind this increased volume which, I believe, is good news and great for the economy. There are huge amounts of infrastructure and healthcare work in the pipeline well into 2030 and it amounts to billions and billions of dollars. This means job demand and security for construction workers at all levels, which begs the question of why there is a labor shortage in the first place.

The other more important factor contributing to the labor shortage was described as a marketing issue in a previous section but will be discussed here as it pertains to DEI. Since many workers in the construction industry are predominantly white males, this has created a schema that other types of people are not suited for a career in construction. From the perspective of an outsider looking in, it has felt like they do not belong or are not welcome.

Now, finding ourselves in a situation where we do not have enough people to complete ongoing work and much more work lined up in the future, why would we limit ourselves to a certain demographic? As discussed earlier, the answer is not to simply hire more diverse people because that alone will only set those workers up for failure and make the problem worse. It is everyone's responsibility to think about the way we think and treat each other to make a fundamental shift in the way the industry has operated for the last few decades. Again, refer to the basics of human connection – trust and respect. The tenet of this concept is respect for all; it is that simple. This may sound drastic and extremely difficult, and you may think that you are only one person, but we all must try. We don't have another option!

It has become extremely difficult to find the right people – or even any people in some instances – to fit certain positions or company requirements. By increasing diversity in our workforce and everyone's acceptance of it as an opportunity, not a threat, we will take huge steps toward solving the labor shortage.

Labor shortage aside, focusing on the key principles that would promote a more diverse and equitable workplace will generally improve the industry. Remember, it all comes down to relationships. So, if everyone takes conscious steps to improve the way they interact and connect with others in a positive and effective way through fostering trust and respect, imagine how much more productive and happier everyone will be. It may sound like I am suggesting that we hold hands and sing songs but allow me to explain a little further.

People want to feel valued, cared for, and respected in the workplace and we fulfill these basic human needs by focusing on the elements of trust and respect through effective relationship building – especially with those who have traditionally been excluded – it is fair to assume that acting in this way will make others happy and fulfilled. A person who is happy and fulfilled will work harder and longer than someone who is mistreated. If the culture starts to develop as more people who are happy and fulfilled start to infect others with their energy, they will start to attract more people who would like to be part of what they have created. All of this will lead to increased productivity which in turn will increase profitability. The result? Happy, productive workers and juicy profits are all from just being kind to people and spending some extra time consciously connecting on a basic human level before adding on the complexities of work and money.

Now it is time for a reality check.

Of course, it is not that easy when factoring in the variables of different personalities and the longstanding generational issues of racism and sexism. I believe that most people are good and truly want the best for others, but they are stuck in the way they think things should be because that is the way they have always done it. It goes back to the previous discussion on change management and fear of the unknown or, in this case, fear that the comfort and predictability in one's life may be uprooted for another's benefit. Sometimes you need to put other people's priorities in front of your own to get ahead. This is true leadership, and it just requires the first step beginning with the conscious choice and effort to connect with respect and care for others.

Sustainability: Beyond Design Principles

Sustainability in the construction industry is often associated with green building practices, carbon reduction, and energy-efficient designs to reduce the impact of buildings and operations on the environment. However, sustainability extends beyond these aspects and permeates the very culture of a company and its network. In this section, we will discuss creating an environment that fosters growth, inclusivity, and longevity for a sustainable career and company.

Sustainability for your career or building your company involves building relationships that are not just beneficial in the short term but are designed to last and evolve over time as your – and others – needs change. It's about understanding that each connection we make is a part of a larger ecosystem, and the health of that ecosystem depends on the strength and quality of its individual components. This means investing time and effort into nurturing these relationships, understanding the needs and goals of the people we connect with, and finding ways to support and uplift each other. In short, it's about playing the long game. For personal connections, this means staying in touch with people and continuing to provide value even when it is past the initial value stage. Using people as a one-time "benefit" will hurt your career and not provide sustainable success. This requires continuous contact and availability to help when called upon, within reason.

In terms of company culture, sustainability means creating an environment that supports the well-being and growth of its employees. It's about recognizing that employees are the most valuable resource and that their happiness and fulfillment have a direct and corresponding impact on the success of the company. The happier your employees → the better they treat your customers → the happier your customers → the more money they give you/the more friends they tell. This is a basic premise but fundamental to understanding the concepts in this book.

The challenge comes in creating a culture of respect and inclusivity, where everyone feels valued and heard. One that provides opportunities for professional development and growth and recognizes and rewards hard work and innovation. Those are all nice words but what do they really mean? In short, it means that employees are valued for who they are, not what they do. Recognition and reward need to be broken down and communicated in a combination of short-term (quantitative) and long-term (qualitative). Short-term rewards (bonuses, same position raises, etc.) are required to incentivize immediate performance but are not enough to align a worker with company values/culture and ensure continued sustainable employment. Most companies are very good at offering short-term incentives (competitive salary, bonus structure, etc.) and this is how many organizations have operated over the last few decades. However, there is a growing trend toward sustainability in careers, focusing on longer-term metrics like tenure, promotions and increased responsibility.

Many organizations fail to identify a career potential path depending on different employees' wants, needs, and aspirations, which is what a lot of younger workers are looking for these days. Yes, there are a lot of people who want to get promoted quickly into senior management and, no, not everyone will, but they all need to see a path on how to get there or they'll go off and make their own. In this sense, communicating career paths includes the how as well as the why to show potential routes and the requirements to travel them. This will provide more of an incentive for longer-minded individuals to stay within the organization.

In conclusion, sustainability is not just about design principles or green building practices. It's a holistic approach that involves every aspect of our work, from the way we build relationships to the way we run our companies. By embracing sustainability in all its forms, we can create a construction industry that is not only successful but also resilient, inclusive, and beneficial to all. Remember, a sustainable construction industry is one that is built to last, focusing on the tasks in front of us while keeping in mind the steps in the future.

Reducing Inequalities

Inequalities in the construction industry are a reality that we must confront head-on. These are some of the biggest elephants in the room when it comes to issues in industry and not acknowledging them is a disservice to everyone involved. They manifest in various forms, from wage disparities to unfair treatment to unequal opportunities for advancement. Reducing these inequalities is not just a moral imperative, it's a strategic necessity for the growth and sustainability of the industry.

Inequalities are caused by several factors – gender (men/women), class (field/office), profession (consultant/constructor), age (young/experienced), etc., are some examples of dichotomies that can create inequalities on site or in the office. Tackling each of these examples would take a book of its own and is not the scope of this work; however, it is important to identify what role you and your company play in addressing the most pressing ones according to your core values.

Company culture plays a pivotal role in either perpetuating or reducing inequalities. A culture that values diversity, inclusivity, and fairness can help level the playing field. It's about creating an environment where everyone, regardless of their background or identity, feels valued, and heard, and has equal opportunities to grow. Leadership sets the tone for the company culture. Leaders who champion equality and inclusivity can inspire their teams to do the same. They can implement policies that promote fair treatment, such as equal pay for equal work, transparent promotion criteria, and flexible work arrangements that accommodate different needs.

Moreover, leaders can foster a culture of learning and development. By providing opportunities for continuous learning, they can ensure that all employees, regardless of their current skill level or position, have the chance to develop and advance in their careers. The key here is breaking down barriers caused by inequality. So, everyone has visibility of the mission and can understand its alignment with core values. It's ok to address the issues without solving them, provided you are acting from a place of genuine care and compassion for your coworkers. Most people will appreciate the authenticity and help work toward a solution. Others will want to maintain the status quo because change is uncomfortable, as previously addressed in this book. It's important to plan how to deal with each side appropriately to keep your vision or career moving forward.

Personal development is another crucial aspect of reducing inequalities. It's about empowering individuals to take charge of their own growth and career trajectory. This involves providing them with the resources, support, and opportunities they need to learn, grow, and excel in their roles. While a lot of this focus is on personal aspects of life such as leadership, productivity, and other soft skills, there are huge transferable benefits to professional lives as well. There are countless apps, books, and programs to help people develop in all aspects of their lives for a minimal cost.

Mentorship programs, as previously discussed, can be a powerful tool for personal development as well. By pairing less experienced workers with seasoned professionals, these programs can provide valuable learning opportunities. They can also help break down barriers and dispel stereotypes by fostering connections between diverse groups of people.

Training programs, both on-the-job and formal education, are another essential component. These programs can equip workers with the skills they need to succeed and advance in their careers. They can also help bridge the gap between different groups of workers by ensuring that everyone has access to the same learning opportunities. There are many online or in-person workshops, seminars, and courses that can equip workers with the tools they need to feel prepared and be successful. Check with your local colleges or construction associations for more information.

Reducing inequalities in the construction industry is a complex task that requires concerted efforts at all levels of organizations. It's about changing company cultures, empowering individuals, and implementing fair and inclusive policies. While the road may be challenging, the rewards – a more diverse, inclusive, and productive industry – are well worth the effort. When we lift others, we rise together.

15

Great People Ask Great Questions

Questions Bring Immense Value

Great questions are an important part of leadership as well as learning. I am a firm believer that there is no such thing as a stupid question, except those that are not asked. Knowing when and how to ask questions is usually learned over time. A lot of questions I find myself asking arise from my own or others' experiences that I have learned from. The intent to ask comes from a desire to avoid past mistakes, or out of genuine curiosity about the event or activity that is going on. This has been a valuable tool that I use daily. If you approach situations with an open mind and ask from a place of curiosity or desire to learn rather than condescendingly challenging people's competence or values, you are much more likely to build the relationship, learn something new or add value to a situation. It all comes down to how you ask the question. For example, if you walk by someone on site who is conducting a task that you have done a thousand times and you notice they are doing something incorrectly, one way of approaching them is, "Hey, what the hell are you doing?" Another more subtle approach may be something like, "Excuse me, I noticed that you are working on this task in this way and I am curious why. I have some experience and may be able to show you a safer/quicker way to do it." I am sure you can tell how each of these statements will be received by the worker. Which one is more likely to have a positive outcome?

The Human Side of Construction: How to Ensure a Successful, Sustainable, and Profitable Career as an AEC Professional, Second Edition. Angelo Suntres.

The nuance lies in how the question is asked. Do focus more on providing value to a person or situation rather than just seeking an answer or opening an opportunity to showcase your skills. This also leaves an option for the other person to make a choice and brings with it care and respect which will go a long way in connecting/developing your relationship. Taking the kind approach will also make it less likely that the other person will react defensively and increase the probability that they will share their expertise with you, either in that moment or in the future. Everyone has something that they can learn or teach someone else.

In some workplace cultures, questions are not welcome or are limited to people from certain pay grades or seniority levels. If you find yourself in a company like this, it is important to stress that your questions are asked with the intention of adding value and not questioning the validity or experience of others. Ideally, you will feel comfortable asking questions freely and this could also be a factor in your consideration or taking/keeping a position.

Well-timed and well-intended questions will always bring value and help align team members; however, the opposite can also be true.

Have you ever been in a meeting or situation where somebody asked a question and the tone in the room immediately plummeted? I am sure we can all recall an experience where a badly timed or poorly intended question sent a conversation or situation into a tailspin. Equally important as asking good questions to bring value is avoiding questions that add no value or even negatively impact people. The best measure of whether or not a question will add value is to pause and think about how you would feel if somebody asked you the same question. What is the intent? If you would not appreciate somebody else questioning you in that manner or if you feel that it comes from a place of personal vendetta, anger, jealousy, or frustration, then it is probably best not to bring it up.

I have seen quite a few meetings turn bad in a hurry because of one bad question that was asked, triggering a tailspin of back-and-forth comments that wasted everybody's time and brought down the collective energy. Remember that time and energy are your two most valuable resources; asking bad or ill-willed questions will not bring any benefit or enjoyment to any situation.

If you are on the receiving end of one of these negative questions, it is always best to avoid snap decisions or snarky replies. This can be hard to do especially in a group situation. When you feel like your credibility or ego is being threatened, know that nothing good will come from stooping to the other person's level. Try to take the high road, stay respectful, quickly shut them down, and change the subject. I find that these situations are best dealt with in private, so request a separate meeting or conversation with the person when it is just the two of you. You can also use your judgement about the level of severity and whether it warrants escalation to a more senior person. While it may feel satisfying putting someone in their place in the moment, consider how this will impact how your team views you as well as what your manager would think if they were there or found out about it through someone else.

When it comes to asking questions, you have the power to choose whether you want to contribute, learn, and add value or slow down, destroy and derail progress. It just takes a moment to stop and think ... choose wisely.

When to Probe and When to Back Off

Through asking questions or observing others asking questions, you will find that some get more pushback than others. Pushback in this sense is defined as the person on the receiving end either not appreciating the question, hesitating to give a response, or being aloof in their response. If the question was well intended and the recipient reacted defensively, it is possible that the question was misunderstood. Note that impact does not always equal intent, meaning that your question may not have been received in the way you intended it but what matters more is acknowledging if it resulted in harm or insult to the other person. If this happens, it is worth apologizing and then clarifying what you asked and why to alleviate any misunderstanding or hard feelings the other person may have experienced.

If you find the second situation is true and they are just not giving a straight answer or avoiding the actual question altogether, then it is time to dig a little deeper. Usually if people are aloof in answering questions, they either are hiding a truth that they don't want people to know about or they simply don't know the answer. If the latter is the case, the simple solution would be acknowledging that they are not sure and that they will look into it, right? For many people, it is not easy to admit that they don't know as they are afraid that it shows weakness and may cause feelings of inadequacy if it is something that they "should" know in their role. If this is the case, refer to your relationship principles of openness, honesty, trust and respect and let the other person know it is okay if they don't know as long as they provide the required information in a timely fashion. If this is the case, the truth usually comes to light pretty quickly with one or two follow-up questions, but the other scenario may take a bit more effort.

If it comes to light that the other person is hiding information or refuses to answer, it may be worth rephrasing the question to ask about more specific items or asking them if there is anyone else who can answer the question for you. They may be hiding their own mistake, or they may be dealing with the decision of a coworker or a more senior person, and unable to answer. If this is the case, you should not push too much because you don't want to put anyone in an uncomfortable situation. How important it is for you to obtain the answer will determine how hard to push in the moment or when to back off and follow up afterward. There is absolutely no benefit for anyone to humiliate somebody for not knowing the answer to a question. Remember the key is to stay respectful of everyone and build connection ... humiliation will do the opposite.

Another important factor to consider is who you are speaking with, and in what forum the question is being asked. If you are in a large meeting with many different stakeholders, people will be less willing to offer up more private information than if it was just an informal chat over the phone. You will likely be able to sense if there is something that the other person wants to say but is hesitating to divulge. If this is the case, you could just simply mention that you will call them after the meeting, or just ask them to follow up with you when it is convenient. It is not advantageous to put them on the spot. You should also consider the experience level and seniority of yourself and the person or people you are talking to. For example, if you are a young engineer speaking with an experienced journeyman or foreman on site, asking questions about a certain material or installation method, it is possible that the other person may not take you seriously or will be offended by your question even if it is perfectly valid. This is not to say you should not raise issues or ask questions, but consider the dynamics before spending too much time or energy with someone who may get agitated and decrease the probability that you will get an answer or make any progress. You may need to find alternate ways to get your question or point across.

When used properly, questions can be a great opportunity to show others that you are critical thinker, add value to situations, and uncover problems that others may have missed, but they could also get you into trouble. Therefore, it is important to consider when, how, and why you are asking your questions and where they are directed to maximize your results. Great leaders have the capability to ask great questions through proper timing, wording, and context even if others perceive this as a weakness or irrelevant. It is easier to be vulnerable and admit you do not know something than to fix a mistake that could have been avoided.

16

Barriers to Human Connection

Technology Is Divisive

You cannot discuss the importance of human connection and communication without addressing the barriers and challenges that must be overcome to achieve them. There are many factors today that slow down or completely block the ability to connect with others. The industry is ever evolving and innovating, a requirement to keep pace with other industries and improve the way we do things, but you cannot neglect basic human factors that have existed forever and will never change. Yes, the forms of communication need to improve to increase speed and efficiency but the communication itself – connection between people – will always be the key.

Through technological advances, there have been many improvements in the speed of communication via emails, text messages, and other forms of instant and direct contact. The construction world relies on email as the main source of communication and document control because it is written and time-stamped, which provides a trackable record and legal document that can be sent to anyone with an email address. As a result, this is an incredible tool for certain purposes such as records of conversations, meeting minutes, shop drawing reviews, RFIs, and SIs. However, context is lost outside of these document-centric examples. Remember that most communication is nonverbal and conveyed through mannerisms, tone of voice, eye movements, etc., which are completely lost in

The Human Side of Construction: How to Ensure a Successful, Sustainable, and Profitable Career as an AEC Professional, Second Edition. Angelo Suntres.

back-and-forth email conversations. Any situation or conversation that requires collaboration, brainstorming, problem-solving, or design development should be done in person whenever possible to allow for full messages across the spectrum of communication from words to actions and energy. Nothing beats being present with another person and being able to look at them in the eye when dealing with a difficult situation or problem you need to overcome. The efforts of collaboration are also increased when you can feel the energy of those around you.

If there is one thing that the COVID-19 pandemic taught us, it is that remote work is possible for most industries and roles; however, the sense of community and human connection is sacrificed causing communication breakdowns and other issues relating to mental and physical health. The bottom line is that humans were meant to connect with each other, which is best illustrated by considering the way we have operated for thousands of years where all aspects of life were based on communities for support, from family to personal issues and work-related instances. Recalling from earlier, "it takes a team to raise a building." Email, Zoom meetings, and other online forms of messaging are great for solving long-distance communication issues but, ironically, while the intent is to bring people closer together, it often results in keeping them apart.

Technology is an important part of the future of construction, especially in the near future where we will face complex projects, shorter schedules, and limited human resources, but you cannot neglect the importance of human interaction in this equation. People will always be the heart and soul of every organization and need to be connected to be at their best.

If you have not picked this up already, I believe that there is a solution to every problem. It just depends on the amount of effort and time you want to spend finding it. It will not come as a surprise to you that, despite the challenges described in the previous section, there are ways to overcome them. The key to overcoming the barriers that technology presents to human connection is identifying how to maximize the effectiveness of your communication. For example, email is a great tool to document something that happened whether it is an event, a daily journal entry, an incident report, etc.; however, if the intent of this is to elicit a response, engage in a conversation, or further a discussion, you should consider a more interactive method such as a phone call, in-person meeting, or a virtual meeting. In my opinion, the level of human interaction required to solve the problem or come to a resolution should inform the type of communication required: the more interaction required, the more physical the means should be.

There is no one-size-fits-all solution, though I have noticed some people seek these out. Often people default to what they are most comfortable with, whether that is talking on the phone, sending an email or chatting over coffee. Again it is important to consider the message you are trying to deliver to determine the best way to deliver it. You could ask yourself the following questions to determine the best format:

- What type of message is it (informational or discussion)?
- What stakeholders need to be involved (one or multiple)?
- What is the preferred style of the intended recipient?

For example, if you are a Project Coordinator and you are trying to describe some design issues from a recent product submittal to your team, it is a good idea to give them advanced notice of the challenges you are facing but ultimately information like this is best described by sending the actual drawing so the person can see it. It is also worth giving some thought as to the type of communicator the person on the other end is, as it may or may not align with your personality or style.

Remember, one of the key themes in this book is stopping to consider the relationship with the other person before conducting your business or delivering your message. This is the best way to add the most value and make the most out of your interactions. While this may seem cumbersome in the beginning, it only takes a few seconds and, when practiced and automated, will be a small investment with a huge return on all of your relationships and your career.

Integrity Is Hard to Find

There is a lack of trust in the construction industry that is prevalent at all levels as well as internal and external to organizations regardless of size and complexity. Owners mistrust contractors, contractors mistrust suppliers and subcontractors, some employers do not trust employees, and vice versa. A great example of the latter is the recent phenomenon of quiet quitting where employers stop going above and beyond in response to not being treated properly or compensated fairly. To clarify, I am an advocate for fair pay and proper treatment as it aligns with my philosophies on treating everyone respectfully; however, neither side of the employment spectrum should give the bare minimum and expect anything more in return. You get out what you put in in life and your career.

Throughout my career, I have met many coworkers, friends, and associates, who have had bad experiences where they were mistreated by others in the past either because of encountering bad people or just being in the wrong place at the wrong time. It seems that this is true for most professionals in the industry, especially those who have been around for several years, but even people who have not experienced this directly have heard stories or been warned which has created an overly defensive environment where people are ready for a fight to protect information and profit margin. The result is an argumentative, combative, and litigious industry and, where these qualities are required in some situations, it should not be this way by default. It is important to learn from your experiences, and as humans, we are programmed to avoid pain and suffering, both mental and physical, but having a constant sense of trepidation and a feeling that someone will do something bad to you at any moment takes away enjoyment and impedes progress.

There are two important takeaways from this section. First, keep in mind that new relationships in the construction industry often start from a negative place where many are expecting you to try to take advantage or be dishonest rather than being neutral at the start. Using the principles of open, honest, and consistent communications and actions, you will be able to make a meaningful connection with most people or at least get to the point where you will have a working relationship. It may just take time. Second, if you have been hurt in the past, I am sorry. This industry is known to be tough. It takes courage, trust, and faith to try again but we all have to do our part to move the industry forward.

Before we can discuss how to regain or repair mistrust that was likely not our wrongdoing, it is important to understand how trust is acquired. In any type of relationship, whether it is business or personal trust, it is the elemental building block that is earned, not inherited. The only way it can really be accelerated is if a person with whom you are building a relationship has been referred to you by a close personal or work acquaintance, but they will still likely need direct exposure over time to build trust with you.

The best and fastest way to build trust is through commonality, consistent effort, showing care, and adding value to the other person.

When you express these qualities to others and if they see mutual benefit, they will connect with you and this connection will help break down the barriers to human interaction. Trust does not necessarily mean that you agree not to share some information about the other person with other people but it is more a sense of safety and security, and that you will operate with the other person's best intentions in mind. If at any point, the other person senses a misalignment between your actions and words, this could negatively impact your trust with the person.

It is also worth mentioning that there are different types of trust in relationships. There are people who may be high performers in organizations. So, they are trusted to get work done properly and on time but would not be trusted with personal information. Conversely, there are people that you may trust with all your secrets, but you cannot count on them to finish a project on time. Simon Sinek put it best in the context of dealing with military affairs when he said, "I trust you with my life, but would I trust you with my money and my wife?"

Think of some examples of people you have met in your life that you know you cannot trust. What was it about them that struck you as odd or made them seem like they were not trustworthy? These are important qualities to keep in when you are looking to connect with others because the qualities that you criticize others are often the ones you are unsure or insecure about within yourself. This may help define or diversify the type of people required in your network and support systems to maximize your success.

Time is Spent Ineffectively

It seems that these days people are a lot busier than in the past, which has impacted how and to what extent we connect. Technology continues to equip us with the power to do more with less, both physical resources and time. An example of technology reducing the need for physical resources is how CAD/BIM replaces drafting tables and revolutionizes drawing and coordination processes. Another example of improvement with respect to time is computers and artificial intelligence being able to generate models not dreamt of 20 years ago. All these achievements are great technological successes intended to streamline and optimize a worker's productivity and I believe that there have been many successes in this regard. However, this comes at a price. The increased reliance on technology has decreased meaningful interaction between people. The efficiencies created with computers and smart phones have clogged up people's schedules with meetings.

Since human connection is dependent on repeated, value-added, and consistent exposure to one another, the fast-paced and busy lifestyle that we are all taking part in makes it more difficult to build meaningful relationships. The result is ironical that we are working with more people but from greater distances and doing more but connecting less. Technology that was designed to bring people together can have the opposite effect when we rely solely on remote solutions. Just consider how much time you spend in front of your computer or smart phone. It is not unrealistic to think that you spend up to six hours on a screen each day and depending on your position, you could spend more. This concept is relatively new to humans from an evolutionary perspective since common use of computers became popular in the 90s and the effects on mental and physical health and wellness are still to be determined.

While there are a lot of powerful tools that exist and continue to be developed, nothing beats connecting with people and building a strong network and support system. The key is to find a balance between both the tech and the people. Technology will likely play a larger role in the industry's future than it does today. There will likely be many more technological advances in the construction industry that will continue to redefine how we build, but people are not going anywhere any time soon either! We will always exist in harmony with technology, and I firmly believe that, no matter how efficient, fast, or strong machines get, they will never match the connection that can be built between two people.

Ignorance Is Not Bliss

In the construction industry, siloing is a huge barrier to human connection, networking, company culture, and personal development. The lack of knowledge or understanding of other people, methods, ideas, and innovation can lead to miscommunication, misunderstanding, and missed opportunities. This has been perpetuated in many cultures with an increased focus on completing one's own scope, schedule, and budget without considering its impact on other stakeholders' interests. With this attitude, while your own work may be protected contractually, the whole project usually suffers as a result.

The construction industry is complex and multifaceted, with various stakeholders involved in every project. From architects and engineers to contractors and suppliers, each party brings unique expertise and perspective to the table. However, if one party is ignorant of the others' roles, responsibilities, or perspectives, it can lead to disconnect and discord. One perfect example is one of the divides addressed earlier in this book, the one that exists between design and construction. Both sides would not exist without the other; however, people on either side can be quick to cast blame on the other because of convenience of "expertise." For instance, a contractor who is ignorant of the architect's design intent may make decisions that compromise the project's aesthetics or functionality. Similarly, an engineer who is unaware of the contractor's practical constraints may design systems that are difficult or costly to implement. These scenarios not only strain relationships but also impact the project's success both from a cost and schedule standpoint, which leaves everyone with an unhappy owner.

Ignorance can also hinder networking and the development of a positive company culture. In an industry where relationships are key, not understanding or appreciating others' roles and contributions can limit one's ability to network effectively. It can also foster a culture of silos and competition rather than collaboration and mutual respect. Furthermore, ignorance can impede personal development. The construction industry is continually evolving, with new technologies, methodologies, and regulations. Professionals who are not proactive in expanding their knowledge and skills risk becoming obsolete. They also miss out on opportunities to contribute more effectively to their teams and projects.

So, how can we overcome this barrier? The answer lies in continuous learning and effective communication in an attempt to understand the perspective and expertise that others bring to the design and build process. Continuous learning involves staying updated with industry trends, technologies, and best practices. It also involves understanding the roles, responsibilities, and perspectives of various stakeholders. This can be achieved through formal education, on-the-job training, industry events, and self-study. Effective communication, on the other hand, involves clearly expressing one's ideas and concerns, actively listening to others, and seeking to understand before being understood. It also involves being open to feedback and willing to admit when one is wrong.

In conclusion, while ignorance may seem like bliss in some situations, it is not so in the construction industry. By promoting continuous learning and effective communication, we can break down the barrier of ignorance, fostering stronger human connections, more effective networking, a positive company culture, and personal development. Knowledge is power, but the real power lies in using that knowledge to connect with others and create positive change.

17

Dealing with Adversity

You Will Get Knocked Down

Despite your intentions and best-laid plans, it is inevitable that you will face adversity at some point in your career for several reasons including people you encounter, decisions you make, and situations that you find yourself in. They may be consequences of your actions, a situation that cannot be avoided or simply just bad luck. Furthermore, there will be people who, for reasons unknown, will not support your ideas or efforts and may even make an active effort to tear you down, regardless of how much time and effort you put into convincing them otherwise or attempting to build a relationship. The world is not always a fair place, and we all have to learn to not take this personally, although that can be difficult at times.

While you should not go into situations or endeavors expecting something bad to happen or planning to face adversity, it is good to consider the possibility of what could come up and what you can do to avoid, manage, or mitigate any potential issues and avoid surprises. Remember that toxic positivity will likely get you into trouble; keep the main objective in mind with unwavering faith that it will be accomplished while at the same time acknowledging and navigating the obstacles along the way.

This is especially important to consider if you are trying something new for yourself or facilitating a change to an existing product or process. When expanding on your current skillset either through increased responsibility or adopting a new skill, you will face a lot of personal adversity as you struggle with the learning curve and making mistakes. When facilitating a change management initiative with a group, you will face many difficulties as people like to cling to what is familiar and comfortable. Refer to the section on change management for ideas on how to navigate these situations.

It is important to keep in mind that every step in your journey is part of how your story will be told and the key is not to get down on yourself or give up because of adversity that you may face. In some cases, it may be self-inflicted through an error or inexperience which, as discussed previously, will provide an opportunity to learn and grow. Many times, the issues that you deal with will be totally out of your control, but you will still have to overcome them to achieve success. You will not be judged by what happens to you, but rather you will be defined by the way you react, adapt, and overcome.

This may be the most important section of the book and critical to your success. Since it is almost guaranteed that you will face some level of adversity in your career, it is imperative to learn how to prepare for and overcome these challenges, so they do not stop you from maximizing your success.

The initial focus should be on attempting to avoid issues in the first place. One strategy to accomplish this is to take a few moments to develop your plan and consider what obstacles may come up at each step and ways to avoid or mitigate any poor outcomes. In many cases, your whole plan will not be clearly laid out as there are many variables along the way that will impact future moves or decisions. If this is true for you, just focus on the next immediate step and repeat the process for every subsequent action. It is worth doing this as a check-in at every major step as well to see if any tweaks or course corrections are required before proceeding. For example, if you are a field engineer with the overall objective of becoming a general superintendent, your main steps would be the positions and associated responsibilities leading up to that, e.g. assistant superintendent, superintendent, etc. Some obstacles that you may encounter in this career progression could include learning about specific trades, scheduling, and team management or dealing with conflict resolution. Perhaps there is an educational component associated with your progression (e.g. GSC or B. Eng.). In any given avenue, you will not be able to capture all the potential issues, but you should at least be prepared for the ones that you are aware of.

We have discussed how relationships play a big role in all aspects of your career and life, so you will not find it as a surprise to hear that your network and support system will have a huge impact on how you deal with adversity. Having an intentional, diverse, and connected network will help you overcome challenges in many ways, some of which are listed below.

1) Coworkers will provide the lateral support required to assist in achieving your goals
2) Friends/Family will provide the emotional support required to keep you motivated and on track
3) Management will provide you with the opportunities to grow and learn
4) Mentors will offer guidance and provide direction

Lastly, it is important to be realistic and keep the end in mind when battling adversity. While the overall goals that you set should be high and push you to your limits, there needs to be realistic steps and enough time for you to achieve them. If you are a project coordinator, who aspires to be promoted to a senior leadership position, it is unreasonable to think this will happen in a year. An idea without steps or timelines is just a dream, impossible to measure, and likely not to be achieved. By breaking the main objective into smaller chunks or sub-objectives, you will avoid frustration and disappointment associated with losing sight of your path and abandoning your goal.

Dealing with Job Loss and Unemployment

Much like the inevitable adversity that you will face on your career path, many will also encounter another situation that is very difficult to experience or even talk about: job loss, firing, restructuring, or layoffs ... whatever you call it, it can happen and when it does, it affects much more than your professional career. These experiences are scientifically linked to the same type of grief that is associated with losing a loved one. They can elicit financial stress and feelings of inadequacy and, while it is not productive to dwell in negative thoughts, it is an important topic to discuss as many will experience it in their lifetime.

If you have recently experienced job loss or are currently unemployed, it is important to acknowledge the cause to help process your feelings and chart next steps. In the situation of layoffs or restructuring, it is likely that you were in the wrong place at the wrong time and does not necessarily reflect on you as a person or worker. In larger companies, unfortunately, it comes down to a numbers game, which may provide a small consolation when it comes to personal concerns, but will not help your financial situation. If your termination was a direct consequence of your actions or inactions, it will likely have a personal impact on top of financial stress and can be extremely difficult to comprehend. Take the time required to fully process this event and address issues you may have. Do not hesitate to reach out to family, friends, coworkers, or other support systems as required to help you sort out your thoughts and determine the best course of action. As we have discussed, try to take mistakes as opportunities to learn and grow even if it is difficult in the beginning.

Remember that your worth is not determined by what other people think of you or how you are treated. Perhaps you were in the wrong environment in the first place or somehow found yourself in a position where your skills were not being utilized to their full extent, which can happen in some cases where job duties change over time. It is impossible to grow to your full potential without the help and support of those around you; so, take the time you need to process your experience, reassess your goals, and find the right place for you to succeed with people who value and care about you. I know that you can do it; the fact that you are reading this book is a huge step in the right direction to maximize your career and the impact you will have on those around you.

The obvious result of experiencing job loss or unemployment is the absence of a paycheck, which can cause a lot of stress. However, there are less tangible and sometimes more significant factors that need to be discussed regarding the impact of this event on your relationships and career goals on a more long-term basis. Losing your job can raise feelings of self-doubt and lower confidence levels resulting in changes regarding how you view yourself as a professional. You may also be self-conscious about what others think of you and worry about explaining your unemployment or loss in your next job hunt/interview. These are all normal thoughts that everyone has in these situations and there are ways to overcome these challenges.

First, it is important to realize that you do not need to lower your goals or expectations of yourself because of a perceived failure. It is likely that you took the right steps but were not in the right setting to facilitate the goal or plan that you were executing. Every experience you have in life is an opportunity to learn; try to use this principle to stay in a positive mindset and continue growing as a person and a professional. Do not lower your standards because of external factors that are out of your control; rather, focus on your circle of influence to develop the skills and relationships you require to move on.

Second, addressing the gap in your resume or explaining job loss is likely not as much of a concern as you may think. The quality of applicants is what hiring managers look for and good leaders understand that being let go from one organization does not necessarily mean that it directly reflects your performance. Remember that relationships play a major role in every aspect of your career. You may have a direct or indirect connection with a hiring manager or other authority in the employer who can personally vouch for you regardless of how your last employment ended.

Last, dealing with other people's opinions and how they perceive you may be the most difficult issue to deal with since it is entirely out of our control in the sense that we have little power over what people do or say and zero power over what they think. Yes, our social networks have a huge impact on our identity, and yes, we have a basic human need to be loved and valued; however, we need to realize that not everyone will have our back and support us especially in times of struggle. Situations like job loss present an opportunity to identify those who do not bring value to your life and allow you to assess the role they will play in your future successes.

The key takeaway here is not to let short-term failure get in the way of long-term success. If you are currently experiencing job loss or unemployment, it is temporary and I wish you the best in finding something sooner than later.

18

Conclusion

This book illustrates how focusing on the human principles of connection – effective communication, fostering healthy relationships, and nurturing strong connected networks – is critical to ensuring a successful future for everyone in the construction industry regardless of role or position. Applying the skills and techniques in this book will have profound effects on both your professional and personal life and will help you stand out from the crowd, build meaningful relationships and projects, and attain success – whatever that means to you.

Whatever your goals are, I truly believe that with the right planning, realistic expectations, and the principles discussed in this book, you can achieve them, no matter how big or how small or how much adversity you face. We are all more powerful than we realize. I sincerely thank you for going on this journey with me and wish you all the very best in your future endeavors.

The Human Side of Construction: How to Ensure a Successful, Sustainable, and Profitable Career as an AEC Professional, Second Edition. Angelo Suntres.

Appendix A

Find Your Calling

Three Questions You Need to Answer

So far, we have covered the importance of connecting with others in the construction industry and hopefully developed an appreciation for the vastness, complexity, and quantity of relationships that you need to establish, nurture, and maintain to achieve success. Before we get into details on how to further connect with others and introduce techniques to build your network and support systems, we will spend some time on the most important relationship of all – the one with yourself. Without a strong understanding of your personal wants, needs, desires, and strengths, it will be extremely difficult to find your place and maximize your success in the workplace.

Depending on where you are in your career, you may or may not believe that this is a relevant section. If you are already well on your way with many years of experience or even in a management role, you may have already figured this part out; however, I believe that checking in with yourself is something that we can never do too frequently. Things change as time goes by and you may realize that your passions and interests differ year over year. The last thing you want to do is continue down a path just because it is what you have been doing and are comfortable with. A cargo ship which sets off for a destination does not just aim at the final location and hope for the best but rather requires regular check-ins

The Human Side of Construction: How to Ensure a Successful, Sustainable, and Profitable Career as an AEC Professional, Second Edition. Angelo Suntres.

to make sure it is on a path. Course corrections may be required depending on factors like weather, which will change the overall path of travel. Your career is no different.

You can make these check-ins as formal or informal as you would like ranging from a simple mental exercise to jotting down notes on a piece of paper or journal. The important point here is to dedicate the time and effort to complete the activity, which should not take more than 10–15 minutes to complete and includes asking yourself the following three questions.

1) What am I good at?
2) What do I enjoy doing?
3) What can I make money at?

In this section, we will elaborate on number 1. Now this may seem like an obvious question to answer but as you develop skills and experience, you might find that there are sub-niches that you excel in over others as you try new things. The key is finding your strengths and weaknesses as well as monitoring them over time so you can use this information to inform future decisions and maximize your success.

For clarity, it is important to define what it means to be good at something. Are there quantifiable results that you can track and measure or is it more of a subjective result in which you may require feedback from others? You might have to take some time to define what this means for you. When I complete these check-ins, I like to look back at my recent accomplishments and see where I received positive feedback from others, whether it was a coworker, manager, direct report, or even my spouse. Sometimes we may think we're doing an awesome job when everybody else thinks otherwise, which is why it is extremely helpful to get feedback from others.

How long you have been in your career will also impact how easily you can answer this question. If you have been in the same field for five years or more, you should have a clear picture of whether you are doing a good job or not. If you have less experience or are just starting out, naturally the learning curve will present challenges. So, it may be difficult to know how you are doing. In this case, you will rely heavily on feedback from others.

It is worth mentioning that there is no limit to the number of items you list in this exercise. If, after careful thought and consideration, you find only one or two items that you really want to focus on, that is perfectly fine. If you end up recording 10 or more items, that is great as well but keep in mind that in future steps, we will be discussing how to take action on these points. So, it is best to focus on your top five to keep things manageable.

The next question we are going to consider is what aspects of your job or day-to-day life do you enjoy doing the most? You do not have to limit any of these ideas to just career-related items as you may be surprised to find commonalities between personal and professional activities. Of course, we all like watching Netflix and eating junk food but those are two extreme examples of something that is probably not going to help you down the road and should not be included in your list. When considering items that you enjoy doing, also consider your "To Do" List. Which items on this list do you look forward to doing or prioritize because you are excited to accomplish them first? This is a good indicator that you find some sort of enjoyment in these tasks.

Another way to determine what tasks you enjoy doing is by assessing how you feel while you are in the act of completing the task. The ultimate example of this would be achieving a state of "flow" where you get so consumed by the activity that time flies by without you realizing it and you essentially get lost in the task. An example of this for me is formatting and compiling data into spreadsheets. I love spending time inputting and formatting data into spreadsheets and find that it is so satisfying that I lose track of time. Remember that you do not have to have an explanation for all of these thoughts and that these notes are private. So, do not worry about what others may think because they will not find out. Even if you think your own ideas are silly or irrelevant, you never know what commonalities you are going to find from this exercise. So, write them down anyway.

You may find that some of these items are the same as the notes that you came up with for the first question. We are not doing anything with this data right now. You just want to record things as they come to mind. Don't worry about trying to find matches, that will come later. Also do not be surprised if the activities you list in this section have nothing in common with the first; just try to focus on writing down as many tasks that come to mind.

Like question #1, there is no right answer and you do not need to list a certain number of points – if you come up with two or three that is fine, if you have ten or more that is great too but you might want to prioritize and find your top five to move on with.

The third and possibly the most difficult question to answer is what you can make money at. The obvious answer here might be the job or career that you are currently employed and/or something similar but feel free to use your imagination a little bit and think outside the box. Depending on the type of company you are at and their philosophy and structure, you might have some flexibility to impact your job description and add certain tasks. This is a great opportunity for you to formulate a plan that you can go to your manager with including ways that you can improve your team or company's performance while increasing your responsibility and job satisfaction. However, in other organizations, this may be frowned upon. So, you have to use your best judgment when approaching your employer. Nonetheless, the purpose of this task is to come up with a personalized development plan. So, let your mind run free and then list out all the ideas that you can think of in which you can make money.

Again, we are not considering previous questions at this point. Right now we are just coming up with all the ideas you can for ways to bring more monetary value into your life. So, for example, if I were to do this exercise, some examples I would come up with would include rules directly related to my experience estimating, project management, site supervision, scheduling, motivation and leadership, and team management. In this exercise, it is not possible to be too general or too specific. So, I would include things like "helping others." If you start broad, it may help you narrow it down. So, from the example of helping others, that may involve mentorships, teaching, coaching, etc.

This should not take too much time or stress you out. You can always stop and revisit it another time too if you are unable to come up with many ideas or any "good" ideas. Like the first two points were not looking for a certain number. So, whatever you come up with is perfectly fine.

Know What You Hate

We all have tasks we hate doing – unfortunately, it is a part of life. In this section, we are going to give some thought to parts of our job or career that we do not enjoy and would rather avoid in the future. Unlike the previous three questions, where we kept it open to both personal and professional items, I would limit this question to strictly professional ones just because if you start listing items like cutting the grass or raking the leaves, that is not going to bring a lot of value.

One way to look at this is to do the opposite of what you did in question two above. When you think of your "To Do" list, which items do you not prioritize, dread getting to, or procrastinate on? This will be your best indicator and maybe even bring to light items that you weren't aware that you did not like. I do not recommend spending too much time on this section because generally I like to keep things positive, plus I am sure that you can come up with a few items pretty quickly. This could include a specific task, dealing with a specific person or a specific trade. Again there is no right or wrong answer and is completely private and up to you personally.

In giving the previous question some thought, we must understand that there are going to be tasks that we do not enjoy doing but we must do anyway. It is worth spending some time here to identify those tasks that you might be dreading but have to be done in order to complete your basic job function or make it to the next level. Some examples of this could include presentations at work if you are anxious about speaking in front of a crowd, addressing a difficult situation that you have been putting off but needs addressed to get resolved, or an upcoming review with the boss, which can be stressful at times, especially if communication is not their strong point with little contact between formal reviews.

Again, to avoid dwelling on the negative, I do not recommend spending too much time on this and likely you will not have a huge list here, but get two or three items that come to mind to help with the exercise.

Assess Your Attributes

Hopefully, that exercise didn't take up too much time, energy, or effort and you were able to record some good physical or mental notes. Now, it is time to look at everything holistically and see how the pieces fit together.

The key element to keep in mind here is to find a way to maximize your enjoyment while making the most amount of money and achieving the most success. Also, we are trying to find a way to complete the tasks or activities that you must do but may not like by exerting the least amount of effort and discomfort. Basically, we're trying to optimize productivity in day-to-day tasks in the present to provide more time and energy to devote the building lasting relationships that will help you advance in the future.

We do this by taking time to review all the notes and see where commonality lies between questions. The ones you really want to focus on are the notes that showed up in all of the first three categories meaning things that you enjoy doing, are good at, and can make you money. These are the ones that you really need to focus on and leverage to maximize success in your career. One question you might be asking yourself is how you can use this information to help your day-to-day activities at work. As we mentioned before, there are aspects of everyone's job that they like and dislike but you have a choice where to invest the most time and energy – if you find that there are activities that meet all three of these criteria, that is a good indication that this is where you should be focusing most of your effort because it is going to bring you the most enjoyment, you are going to do the best job and see the greatest return. Unfortunately, that does not mean that you can ignore other tasks, but you just choose not to prioritize them or put as much energy or effort into them, rather than just getting it done.

You might find that there are items that you enjoy doing and that you are good at them, but they do not present the opportunity to help bring monetary value into your life. Obviously, if they bring you enjoyment and you are good at them, you will want to continue doing them to help feel fulfilled, but with the understanding that there is no financial reward – or at least not in the short term. This may impact your decision to spend time and effort on these activities. We all need to be happy but depending on your financial situation, this may not be an area to focus on above others.

Another combination are tasks that you are good at and can make money in but do not bring you joy. This combination might resonate most with those who have been in the same position or career for quite some time and have developed the skills to do great work efficiently but do not get the most pleasure out of it. If you find a lot of activities in this section, it may be time to consider some form of career change. This does not have to be a drastic case where you approach your manager on Monday with the resignation letter but rather an opportunity to explore new tasks or responsibilities outside your current role or job description. Not only will this give you an opportunity to learn and grow in a new area, but it will show your manager that you are keen on taking on new tasks, growing as an individual, and helping out the team or company in additional ways.

The last combination to consider are items that you enjoy doing and can make money at but you have not yet developed the skill set to be efficient in. The important part about identifying tasks or activities in this area is that there is an opportunity to invest some time and effort into personal or professional development to get the skills required. Again, this may warrant a conversation with your manager to see if there are any courses, books, or other resources available to help you grow in this area. Just like the last point, this conversation will show initiative and will likely strengthen your relationship with your manager and help you stand out from the crowd.

Now that you have taken the time to review personal attributes – what you are good at, what you enjoy doing, and what you can make money at, you should have a clear picture of what type of role you would find the most success in. The next step is to consider how these personal attributes fit into the team setting. So, we will give some thought as to what type of environment will help you thrive.

Appendix B

Find Your Place

Who Inspires You?

Now that you have identified what you can and need to do personally to succeed, it is time to consider the best environment for you to apply it. The first things I suggest looking at are the types of people that have helped you succeed in the past or can help leverage your skills. It is probably easiest to do this by starting with a specific person, who has helped you in the past. Once you have this person in mind, start to dig a little deeper to determine exactly how they helped you and what parts of your relationship with them brought you fulfillment, satisfaction, or any other positive feelings.

One example of this for me is a former coworker, Jim, with whom I had the pleasure of working for three years on a project about five years into my career. At the time, I could not explain why, I just enjoyed his company and had an immense amount of respect for him. After giving it some thought, I was able to break it down into three key areas:

1) He was my mentor. He always took time to explain things to me. He never made me feel stupid to ask any questions.
2) He gave me the freedom to make mistakes and learn the hard way. He was a hands-off leader that gave me free rein to try new things and learn some things the hard way even when he knew better.
3) He always had my best interest in mind even when we were dealing with a problem I had caused or a mistake I made.

The Human Side of Construction: How to Ensure a Successful, Sustainable, and Profitable Career as an AEC Professional, Second Edition. Angelo Suntres.

In this example, I was able to take what seemed like a feeling at the time and then break it down into the attributes that I could use for my personal and professional development.

By doing this exercise, you can create an avatar of the type of people that you should seek to spend your time with, whenever possible. Unfortunately, we rarely get to choose our bosses and coworkers, but you should try to seek out the people who lift you up even if it is just a form of peer-to-peer relationships, mentorships, or friendships outside of work. You never know how these links can benefit you in the future.

The focus here should be on creating the best environment for you to leverage your skills and improve on your weaknesses to maximize your probability of success.

If you are in between jobs or looking to make a change, it is important to consider the type of company that you are looking for. This could be defined in different ways but ultimately comes down to culture, philosophies, and core values. This can be difficult to determine with larger organizations as one department or team may not represent the overall thoughts and beliefs of the company as a whole. This is yet another reason how it all comes back to people in relationships – you could have an outstanding organization with top-notch management but if one weak or toxic leader manages to slip into the ranks, it can cause a poor experience for employees and/or customers and leave a lasting effect. Therefore, it is important to keep in mind not only the organization but who your manager and teammates would be, if possible.

If you have applied and are interviewing for a position, you will likely have the opportunity to spend some time with the hiring manager, which can provide you with a good indication as to the type of person they are and how aligned they are with the overall company philosophies. There are many tools that you can use to do some research on companies and individuals within them. A quick Google search can tell you also a lot about a person and other services like LinkedIn will show common contacts that you may be able to reach out to in order to get an unbiased opinion on the company or person. If you are considering making a career change or switching companies, I highly recommend getting to know the atmosphere you are potentially getting yourself into and making sure it aligns with your personal strengths and weaknesses rather than just focusing on high-salary benefits or other types of compensation. Ask anyone you know who has ended up with a toxic leader or in a toxic work culture and they will likely tell you the added stress aggravation and anxiety of dealing with people or day-to-day situations is not worth even a high-ticket salary. Everyone has their limits and it is important to know yours, for both your daily survival and your future aspirations.

Who You Should Avoid

Equally as important as finding the type of people who lift you up is identifying the ones who bring you down. For a lot of people, it is easier to come up with ideas in this section than it was in the previous one but if it requires some thought, that is okay. Like the example above, the opposite could be true. Someone could come to mind that you just did not click with or just seems to rub you the wrong way. If this resonates with you, give it some thought and see if you can label what exactly about this person you did not like, why you did not get along or why they were difficult to work with.

This could also include types of leadership that you have not worked well with in the past such as a micromanager – I have yet to meet anybody who appreciates working with a micromanager but for some reason, they still exist all over the place, probably because they are extremely gifted at doing the work one level below where they were promoted to. The most difficult part of leadership for most high performers is letting go of control and having faith in others to complete their work, this is also one of the most important aspects of leadership.

This consideration should not just be limited to your manager. Because you spend the majority of your workday with coworkers, it is very important to determine the type of people that you cannot tolerate and do what you can to avoid them in your current or next career adventure.

Like above, if you are in between jobs or considering a change, it is important to consider company attributes that you have not meshed with in the past to avoid getting into similar situations in the future. For example, if your current or old company promoted hustle culture and made it the norm to work additional hours and go beyond reasonable expectations, you may want to avoid getting into the same mess on your next endeavor. Conversely, you may have found that the past or current employers are too laid back, entrepreneurial or free-spirited in providing guidance and you wish to seek a little more direction. In this exercise, it is equally important to determine what does work for you and what does not work for you to ensure that you find the right environment to maximize your results and overall success.

Appendix C

Let's Get Real

What Do You Want?

Before you can sharpen the saw, you need to know what tools you are working with! In the previous chapter, we covered personal attributes and higher-level purpose/enjoyment/reward-fueled ideas. In this section, we will dive deeper into determining the specific tools you will require to accomplish your goals. For example, if you are currently a Project Manager who is looking to gain more responsibility or promotion to senior PM or even construction manager, even if you are a master of the people and relationship side of the business, you still must master the basic requirements of the role. I have always told people that you would be amazed how far you can get in life just by completing the bare minimum very well, on time, and getting along with others. Most people struggle with even finding two of the three items in that formula.

So, let's continue with the example of a project manager who is looking for growth and has identified in the previous chapter that they are good, enjoy, and can make money at negotiating contracts. Having the skill, enjoyment, and potential for reward alone does not mean you are going to be successful. In this example, the project manager may need to hone their skills in specific parts of contract management, e.g. they may excel at negotiating price but have neglected something such as scope delineation or schedule accountability, or maybe they excel in certain trades (structural steel) and could use improvement on others (mechanical/electrical). Another

The Human Side of Construction: How to Ensure a Successful, Sustainable, and Profitable Career as an AEC Professional, Second Edition. Angelo Suntres.

great example is improving knowledge and efficiency in common software. If you have been in the industry long enough, you will know that the world runs on spreadsheets. Mastering the art of Excel or Google Sheets is a perfect example of a skill that anyone can improve on to help accelerate their career from simple tables to pivot tables/charts and graphs. I can think of numerous times where spending an hour or two on learning a new Excel function or chart/graph has wowed many people in management.

Again, now is a good time to get another piece of paper and jot down some job-specific skills that you will require in order to become extremely efficient in your current role – and possibly stretch into others – which will help you get noticed and move ahead. You don't have to spend too much time on it, and you probably don't need to look too far outside of your daily routine.

What repetitive tasks do you do daily? Are there improvements you can make to these processes to be more effective with your time/your team's time?

What software do you use? For estimating, it could be Planswift, AccuBid, AutoBid, etc. For project management, it could be Oracle, Asana, Jonas, etc. Focus on the programs and processes that are already in place and master them. By getting to know these systems, you may even find ways to improve upon them, which may be a welcome suggestion to your manager.

The key to this section is taking a step back and analyzing the critical skills that you need to focus on to excel in your current role. Once you can identify a list of these skills, we can move on to the next section, where we will learn how to leverage them to make you the go-to person and set yourself up for growth.

In a previous section, we covered how to identify what skills are required to do your current job really well. Once you have identified these skills and started to improve your knowledge in the area either through specific courses or training or just your own on YouTube (which can be very effective these days), you just start doing your job really well and people will notice you and get promoted, right? Well, I am sorry to say that this is the point where you have to combine a little bit of art with science. Working on improving your technical skills alone will not unfortunately translate into immediate success. You must leverage these skills in a certain way that the right people notice.

The last thing you want is all your efforts and genius in a specific area go unnoticed or even worse having your credit taken by a coworker or a boss with selfish intentions.

Now just as a warning, there is a very fine line between showboating and letting everyone know that you are really good at your job. The key is to demonstrate your ability or abilities in certain areas not just use words to tell people how awesome you are. You do not want to come across as overbearing or trying too hard because this will work against you regardless of how technically skilled you are. An example of what not to do in this situation is getting into a conversation with a coworker or manager and expressing that you are really good at this certain task. You may be the best in your field but until you demonstrate it, words mean nothing.

The most challenging aspect here is getting in front of the right people the right way. The right people do not necessarily mean your direct managers or other members of leadership, using these skills can build a lot of rapport with your coworkers and may help build the following of supporters that will help you in the future.

Even though you may have spent lots of time and effort learning the best and the most efficient way of completing a task, we should always be open to new ways of doing things. A great technique to achieve this is simply asking a coworker how they go about doing a certain task and can create two different scenarios. One, you could find that there were steps that your coworker was doing that you were not – now you could use them to improve your tasks. Two, you may find that they are doing too many or missing steps at which point – if it is appropriate – you could offer suggestions on how to improve or simply explain how you do it. The latter example would be less confrontational, whereby you provide information on how you do it. Then give them the choice of whether they will accept it. By doing this, you may have improved the way you do things or the way your teammate does things, either way, there was some sort of improvement and progress made for the team. You could then take this information to your manager and say that you have reviewed the process with your teammates and between the two of you came up with some ideas on how things might be able to be improved or just comment that the team is doing a really great job and they should be proud. This shows initiative and care that you have for the team and company and will leave a positive impression on your manager.

The key takeaway from this chapter is to use the skills that you identified in the previous section, learn how to do them well, and demonstrate to your team and leadership that you are a high performer. By focusing on certain skills and not spending time on less important peripheral tasks, you will easily distinguish yourself from others and receive positive attention.

It is always good to focus on positive aspects and things that you do really well but equally important to identify and enhance or support your weaknesses, which we will cover in the next section.

What Is Holding You Back?

In my experience and many others, I have spoken with over my career, acknowledging mistakes and weaknesses is the most difficult obstacle to overcome when it comes to personal development – I think it is human nature to cover up our weaknesses for fear that we will be judged or ridiculed because of them. It is impossible to become a great leader without acknowledging that there are skills or attributes that are outside our zone of expertise and others are needed to complement and support our qualities. This is especially true in an industry like construction where there is a large amount of technical information and collaboration is critical. No one is perfect, and we all have areas of our life that we could use improvement on, but we do not always know what they are – I am sure others could tell you pretty quickly what yours are! The sooner you realize and acknowledge these blind spots or weaknesses, the sooner you will start to implement improvements and move toward maximizing your success.

Like previous sections, you may find it beneficial to take a few moments and jot down some weaknesses that come to mind. You may have to think a little bit about this one, but it is important again to acknowledge that it is perfectly normal to have areas we need to improve on and remember that you do not have to share this list with anyone – be honest with yourself to get the best results.

One example of a weakness that comes to mind when I do this exercise is a lack of patience in the decision-making process. When it comes to making decisions, I am very intuitive and rely mostly on gut feelings after a high-level review of data rather than pouring over details. While this can be beneficial in some situations, I need to remind myself to take time and properly analyze data and get input from others to ensure the best possible decision.

Another common example that holds a lot of people back is perfectionism. This could result in a long painstaking meticulous process to ensure that no steps are missed, revisiting, and tweaking a product. So, it is never quite finished, or scrapping an idea and starting from the beginning with some difficulty or mistakes were encountered. Perfection is the enemy of complete. Sometimes good enough is good enough especially in construction, that is what tolerances are for.

Again, it is important not to spend too much time dwelling on negative topics like this, but chances are after giving it some thought, you can identify at least one or two areas that you need to improve upon. If you end up with five or more, try to focus on two or three at a time as more than that could be overwhelming. You should be able to prioritize in order of importance and start with the biggest ones first. The next step is giving some thought as to how we can take these weaknesses and create an opportunity for improvement.

After completing the last section, hopefully, you are not too down on yourself – the intent is to identify area weakness not to focus on negativity but to find ways to fill or compensate gaps in our skill sets and attributes to complement our strengths. In this section, we will discuss how to turn these weaknesses into opportunities to improve your personal and professional experiences.

Chances are at this point, some of the areas that you have uncovered may require some external help to solve by way of courses, help from peers, or support from family and friends. Whatever is required, know that there is nothing you cannot overcome with a little bit of time, dedication, and effort. Regarding the example I gave above about my lack of patience and decision making, I was able to do some research on my own via books and online videos about analytical decision making to help me improve. As for the other example given above regarding perfectionism, a good measurement here might be to compare your work to that of your peers or some other industry average. There is nothing wrong with completing a task less than perfect and acknowledging that by human nature, no one is perfect. This will involve some difficulty in being vulnerable when you do open yourself up for criticism but if you are in the right environment surrounded by the right people, this will not be a major issue and with the input of others, you will be able to get closer to perfect than you did on your own.

Give some thought to the items you listed and come up with some ideas on how you can strengthen your skill set or attributes in these areas. Now things are starting to take shape!

In the next section, we will cover how we can take all the information previously reviewed and discussed and put it into an actionable plan that will help you all through your development journey.

Make It Happen

By this point, you should have an idea or written notes on the following areas:

1) Personal Attributes:
   - What you are good at
   - What you enjoy doing
   - What you can make money at
   - What you dislike doing
   - Where you can thrive
2) Professional Skills:
   - What skills do you need
   - How to leverage these skills
   - Weaknesses/Blind spots
   - How to improve on weaknesses

Now is a good time to give some more thought into all of these areas to see if there are any particular items you missed or may have changed your mind on. Once your ideas are complete, it is good to look at all of this information holistically to see how it fits together. You may find consistencies or inconsistencies with your responses that will enable you to conduct some deeper thinking into attributes or skills that you would like to develop or not to assist in your current role. You may find that your answers align you with something like what you are currently doing but with a different capacity.

An example of this is if you are a Field Engineer but find that you enjoy, excel, and can make more money in the office as a Project Manager, you may consider what would be involved in making this change or bringing alignment to your strengths and values, e.g. learning contract law, cost control, etc. At some points, you may feel that the answers to all these questions are pointing you in a completely different direction than the way you have been going – now is not the time to make any drastic or hasty decisions but consider all this information as it sheds light on where you would like your career to go.

The reality is that many of us feel stuck in our current roles and are bound by obligations, whether financial, family, or otherwise. We simply cannot all instantly transform our careers into our dream jobs but completing this exercise takes you one step closer to at least developing a plan. Like any good plan, it should be time-bound and include steps to reach the goal, the number of steps, and durations will depend on the size and scope of the achievement. For example, if you are currently a Project Coordinator and are looking to get promoted to Project Manager, it is realistic to think that this can be accomplished in one to three years depending on the size of projects that you are currently working on and the opportunities that exist within your organization. On the other hand, if you are currently an entry-level Estimator and your goal is to be Chief Estimator, this will take some more time and require a lot of experience and education – be realistic and give yourself grace to make mistakes and change the plan when required. The last thing you want to do is set unrealistic expectations for yourself, which will just cause disappointment and frustration and make it more likely that you will give up before you achieve the success you are looking for.

Index

The Human Side of Construction: How to Ensure a Successful, Sustainable, and Profitable Career as an AEC Professional, Second Edition. Angelo Suntres.

d